The China Circle

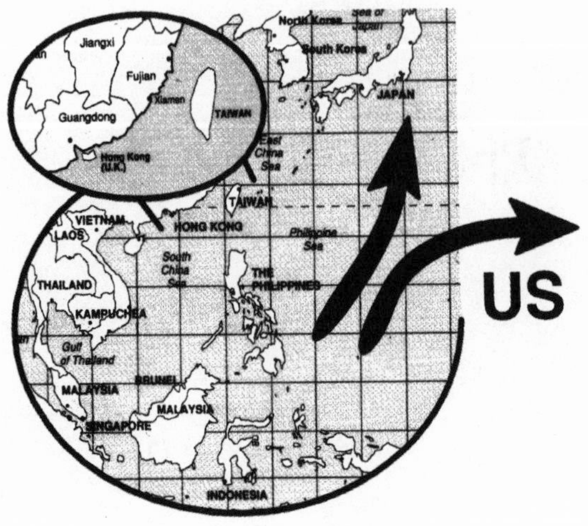

The China Circle, centered on Hong Kong

This book is a project of the University of California
Institute on Global Conflict and Cooperation

BARRY NAUGHTON
Editor

The China Circle

Economics and Electronics in the PRC, Taiwan, and Hong Kong

BROOKINGS INSTITUTION PRESS
Washington, D.C.

Published by Brookings Institution Press
1775 Massachusetts Avenue, N.W., Washington, D.C. 20036

Library of Congress Cataloging-in-Publication data

The China circle : economics and electronics in the PRC, Taiwan, and
Hong Kong / Barry Naughton, editor.
p. cm.
Includes bibliographical references and index.
ISBN 0-8157-5998-3 (cloth: alk. paper). -- ISBN 0-8157-5999-1 (pbk. : alk. paper)
1. China--Economic conditions--1976- 2. China--Economic policy--1976- 3. Taiwan--Economic conditions. 4. Taiwan--Economic policy. 5. Hong Kong--Economic conditions. 6. Hong Kong--Economic policy. 7. Electronic industries--China. 8. Electronic industries--Taiwan. 9. Electronic industries--Hong Kong. I. Naughton, Barry.
HC427.92.C4637 1997
330.951'05--dc21 97-21095
CIP

9 8 7 6 5 4 3 2 1

The paper used in this publication meets the minimum requirements of American National Standard for Information Sciences—Permanence of Paper for Printed Library Materials, ANSI Z39.48-1984

Typeset in Times

Composition by Cynthia Stock
Silver Spring, Maryland

Printed by R. R. Donnelley & Sons Co.
Harrisonburg, Virginia

Preface

In the past fifteen years the People's Republic of China, Taiwan, and Hong Kong have become one integrated economic region. Often referred to as "Greater China" (a term that is offensive to some of its Southeast Asian neighbors), this region, here called the "China Circle," is driving the dramatic growth of investment, production, and trade in East Asia.

The book tells how the China Circle emerged, stimulated by economic complementaries, once political barriers were lowered. Firms, especially small and medium-size family firms, play central roles in the story, with government policies playing secondary, reactive roles. In contrast to the Japanese subcontracting networks in the region, which are tightly organized, the Chinese firms are linked in open, flexible networks, particularly in the electronics industry, and these networks are often led by U.S. companies. In fact, the authors find surprisingly broad and deep economic links between the United States and the China Circle that benefit both. The thickening of economic ties has not eliminated the political rivalry between the PRC and Taiwan, nor has it eased the political uncertainties over the retrocession of Hong Kong. Looking at the China Circle in a regional context, the authors also show that this new powerhouse presents a formidable challenge to Southeast Asia and Japan. The book argues that for the PRC to capture a larger share of the economic benefits from its business ties with Taiwan and Hong Kong it must deepen market reform. It concludes by projecting that the opening of the mainland market could "serve as the engine for a new and historically unprecedented level of Asian economic integration."

The chapters in this book originated from an international project of the University of California Institute on Global Conflict and Cooperation (IGCC) on the impact of the evolving economic ties within the China Circle on the Asia-Pacific region as a whole. The project received financial support from the Ford Foundation and Rockefeller Brothers Foundation. The editor of the volume, Professor Barry Naughton of the University of California, San Diego, Graduate School of International Relations and Pacific Studies, joined me as co-leader of the project. He was ably assisted in the preparation of the manuscript by the excellent IGCC staff, Jennifer Pournelle, chief editor, and Michael Stankiewicz, policy researcher for Asia, in particular.

We especially appreciate the unflagging suport given this volume through its several revisions by Nancy Davidson at the Brookings Institution Press; for helpful comments on the earlier drafts from Nicholas R. Lardy, Terry McGee, and an anonymous reviewer, and for Joni Harlan's and Caroline Lalire's painstaking copyediting. All were instrumental in ensuring that the book become significantly more than the sum of its chapters. And finally, our thanks to the other members of the Press's staff who worked on the book—Jill Bernstein, Deborah Styles, and Susan Woollen—and to Carlotta Ribar for proofreading the book and to Julia Petrakis for preparing the index.

Susan L. Shirk, Director, IGCC
June 1997
La Jolla, California

Contents

Tables

Figures

INTRODUCTION

CHAPTER ONE

The Emergence of the China Circle

Barry Naughton

THE EMERGENCE of an economic region encompassing Taiwan, Hong Kong, and the People's Republic of China (PRC) represents in certain respects the triumph of economics over politics. Three very different political entities have today become an economic trading and producing region, despite political and military conflict and long-standing, deep-seated suspicion and mistrust. No major bilateral or multilateral agreements mark the development of economic ties. Individuals, driven by their own self-interest, acting without government sanction and sometimes in defiance of government controls and restrictions, have created impressive cross-border networks of firms and of production and trade. Economics is clearly the engine driving these three entities into an increasingly intimate relationship.

A closer look, however, reveals that evolving government policy has also been an essential part of the story. Modifications of long-standing policies have been crucial in accommodating the emergence of economic ties. Indeed, in some respects governments have been almost as entrepreneurial as private businesses in facilitating and stimulating lucrative economic activity. Like firms, governments can be seen to be reacting to changing economic circumstances that present new challenges and opportunities. Moreover, as the dramatic events in the Taiwan Strait in 1996 show, economic ties continue to unfold within the context of a ferocious rivalry between the PRC and Taiwan, a fierce struggle for legitimacy and economic and political maneuvering space that can escalate on occasion into overt military confrontation.

The unusual relationship between government policy and private action in the formation of this economic region requires a departure from the standard approach to the material in this volume. The traditional approach is to consider an individual economy, governed by its own policy regime. As that economy develops trade and investment links with other economies, the "international" dimension becomes an increasingly important part of the analysis of the "domestic" economy and domestic economic policy. This approach, by itself, is clearly inadequate for the China Circle. Firms in all the China Circle economies are deeply involved in cross-border networks, and many have intimate knowledge of potential partners and locations in other jurisdictions. To a certain extent, China Circle firms are all located in a common economic region that has been divided by political boundaries. As economies grew and politics receded, the potential significance of this hibernating economic region grew substantially, and when constraints were removed, interactions among its constituent parts grew explosively. Changing capabilities and competitive environments have driven changes in both firm decisionmaking and government policymaking.

The objective of this volume is to describe and analyze the emergence of the China Circle economic region both from the perspective of government policy and from the perspective of firm decisionmaking. In the first part of the book, comprising three chapters, the emergence of the China Circle is examined from a more or less traditional perspective, suitably modified to incorporate the special characteristics of the Circle. These chapters look primarily at the evolution of "national" economies and the modification of government economic policies that corresponds to different stages of evolution. In the second part of the book, four chapters examine a specific industry, electronics, within the context of the emergence of the China Circle and of the world electronics industry. The individual industry case study provides concreteness and, more important, enables us to shift perspective. The dramatic changes within China Circle economies are seen to be driven to a significant extent by conditions outside the Circle: transnational production networks (especially Japanese, U.S., and "Chinese" ones), changing technological capabilities, and changing product markets and competitive conditions. The emergence of the China Circle cannot be understood without reference to these external changes. The volume concludes with a brief assessment of the future direction of the China Circle after its recent explosive growth.

What Is the China Circle?

There is not a single China Circle but rather a series of concentric economic circles, centered on Hong Kong. As Yun-wing Sung (chapter 2) points out, Hong Kong's multiple economic functions give it unambiguous centrality no matter which definition of the China Circle is used. For that reason, the careful analysis and untangling of Hong Kong's economic data carried out by Sung is the essential prerequisite for understanding the economic China Circle as a whole. Even the smallest of the China circles spills across political boundaries, and the larger ones encompass many different political jurisdictions. The fundamental characteristic of the China Circle is the existence of economic ties that cut across political boundaries. The recent growth of economic activity has tended to recreate "natural" economic regions that until recently had seemed to be permanently sundered by political divisions.[1]

The smallest circle encompasses the greater Hong Kong metropolitan area. As the Hong Kong economy has grown, it has extended farther out and become more geographically differentiated. Business services, including finance, advertising, and communications as well as most corporate headquarters, remain in the central business district, whereas most manufacturing has been pushed toward the suburban periphery. This process is an ordinary part of urban growth, but because Hong Kong's periphery is located inside the boundaries of the People's Republic of China, the process has had international ramifications. By now most manufacturing activities carried out by Hong Kong businesses have been relocated to the Pearl River delta hinterland in Guangdong province.

The next largest circle comprises Taiwan, Hong Kong, and the southern coastal provinces of Guangdong and Fujian. Large flows of trade and investment reflect the increasingly complex division of labor among these regions, which are now characterized by production chains that routinely extend across political boundaries. Trade of intermediate goods among production sites accounts for a large proportion of total trade in this circle. It now clearly constitutes an economic region in the sense that manufacturing production networks are highly integrated throughout the area. In other respects, though, there are still important limits on economic integration. There are tight controls on labor movement, and capital flows are subject to many obstacles. Tariffs are significant (though generally waived in the case of materials and components imported for export production), and exchange risk remains. Even after 1997, when it returns to Chinese sovereignty, Hong Kong will maintain a separate customs territory, a separate currency, and controls on migration.

The third concentric circle—perhaps best thought of as an economic region in the process of formation—incorporates all the Chinese economies within the framework of the greater East Asian region. Investment from Hong Kong and Taiwan was never confined solely to Guangdong and Fujian and is spreading northward along the China mainland coast. Investment from Japan, and recently South Korea, initially localized in north coastal Shandong and Liaoning provinces, is growing and spreading southward. These two movements along the coast meet and overlap in the lower Yangtze provinces of Shanghai, Jiangsu, and Zhejiang, the area of China that has traditionally been the wealthiest and most developed. The lower Yangtze provinces have recently been enjoying renewed central government attention, and all are now growing rapidly and attracting substantial foreign investment. As a result of these trends, the entire Chinese coastal area is becoming increasingly open to the world economy, and the economic gap between coastal and inland provinces is widening. Increasing trade and investment flows link China with Southeast Asian countries, and these flows are often managed by ethnically Chinese businesses. Tan Kong Yam (chapter 4) shows that Hong Kong and Taiwan are already among the most important foreign investors in Thailand, Malaysia, and Indonesia, and Taiwan plays an even more prominent role in Vietnam and the Philippines. Thus the same economic forces that are leading to closer links among Hong Kong, Taiwan, and the PRC are leading as well to closer links among Hong Kong, Taiwan, and the Southeast Asian economies. Indeed, as Tan describes, the result has been both competition and mutual benefit in economic relations between China and ASEAN (Association of Southeast Asian Nations) countries, especially Malaysia, Indonesia, and Thailand. Although it is premature to claim that these increasingly dense transactions networks define an economic region, they certainly point to a future in which all coastal China is integrated into the larger East Asian and world economies. Hong Kong and Taiwan will play a critical role in this integration process.

No one of these definitions of the China Circle is the single correct one for economic purposes; the appropriate definition depends on the question being asked. However, the second circle—Hong Kong, Guangdong, Fujian, and Taiwan—represents the most striking and surprising phenomenon: close integration of production networks *without* any institutions for fostering overall economic integration and *despite* significant unresolved political differences. This most significant "China Circle" has grown extremely rapidly in recent years. In 1985 the sum of exports from Taiwan, Hong Kong, and the PRC provinces of Guangdong and Fujian amounted to 2.8 percent of world exports; by 1995, they were 6.6 percent of world exports.[2] Even taking into consideration the fact that much of the increase in exports is due to increasingly dense trade links among the three jurisdictions, this increase in the share of world trade activity is remarkable. Moreover, in the process, the region has emerged as an important center of global production in a wide range of industries, beginning with labor-intensive industries such as toys and garments. In the computer industry, this China Circle is the third most important producer and exporter in the world, following only the United States and Japan.

Rapidly increasing trade and investment have transformed the provinces of Guangdong and Fujian in a manner quite different from their impact on the rest of China. Foreign direct investment (FDI) into China has been very large since 1993. According to official data, realized FDI in the PRC grew gradually from nothing to about $5 billion[*] in 1991, and then surged to $28 billion in 1993. Since 1993 the inflow has continued to grow, running at 4–5 percent of Chinese gross domestic product (GDP), and surpassed $40 billion in 1996. However, FDI into Guangdong and Fujian in all periods has been nearly five times as important, relative to GDP, as in the rest of China. This means that since 1993 annual inflows have been close to 20 percent of GDP annually for these two provinces. A similar qualitative difference exists for trade: in 1994 exports were 106 percent of GDP for Guangdong and 30 percent of GDP for Fujian, compared with 13 percent for China

[*] All dollar amounts are U.S. dollars unless otherwise indicated.

excluding Guangdong and Fujian.[3] It is arguable that these two provinces are more closely integrated with Hong Kong and Taiwan than with the rest of the China mainland, at least judging by commodity flows.[4]

Hong Kong has been the intermediary facilitating the integration of these Chinese provinces into the world economy. Moreover, the efficiency of Hong Kong and the liberalization of the economy in Guangdong and Fujian have attracted trade and production into the China Circle from the rest of the PRC. As a result, the center of China's international trade economy has shifted south to Hong Kong. In chapter 2, Sung's analysis of the Hong Kong trade and investment data shows that Hong Kong's importance to PRC trade has increased in both absolute and relative terms. Moreover, Sung uses the data to describe and differentiate Hong Kong's role as entrepôt, as investor, and as trading partner. The political incorporation of Hong Kong into the People's Republic on July 1, 1997, will not change any of the fundamental economic factors that have led to Hong Kong's emergence as the economic capital of a south China region, but new forms of political risk will arise.

Changes in the External Economic Environment

The emergence of the China Circle came in the wake of three changes in the fundamental economic environment confronting the Chinese economies. These three changes affected all developing economies but had a particularly immediate impact on the China Circle. The first was the increasingly obvious success of export-oriented industrialization in Japan and the East Asian newly industrializing economies (NIEs). The second was the reduction in transaction and transport costs that made it possible to relocate stages in production chains to low-wage sites. The third was the collapse of raw material prices in the early and middle 1980s that made natural resource-based development strategies increasingly unattractive. Realignment of the East Asian currencies—starting with the appreciation of the Japanese yen in 1985 following the Plaza Accord—then provided the catalyst for rapid economic change in response to changing environmental conditions. In turn, these environmental changes provided the external context for the deepening process of economic reform in the People's

Republic of China, which was itself an indispensable precondition for the emergence of the China Circle.

The basis for the emergence of the China Circle was the success of Taiwan and Hong Kong in developing labor-intensive manufactured exports during the 1960s and 1970s, particularly to the U.S. market. This success had both a demonstration effect (as China sought to emulate their success) and a restructuring effect (as export surpluses, increasing costs, and currency realignments created strong incentives to move production to lower-wage locations). These trends reached a high point following the currency realignments of the mid-1980s. The Japanese yen appreciated sharply between 1985 and 1988, and the Taiwan dollar followed the yen upward after 1986. In response to rapidly increasing wage rates and labor shortages, as well as to increasing costs for land and environmental protection, businesses in Hong Kong and Taiwan (as well as in Japan) sought to restructure existing export production networks.The opening of China to foreign investment at this time created a dramatic opportunity to transfer labor-intensive export production to the People's Republic. Wages were much lower, and land and operating costs were lower as well. In chapter 3, I describe the interaction between economic changes and policy developments in both Taiwan and the People's Republic. Policy changes in both locations were driven by these economic changes: the PRC responded to a perceived opportunity, while Taiwan responded both to the pressures to restructure created by earlier success and to the new opportunities presented by the mainland. In chapter 4 Tan stresses that these forces were at work throughout East Asia and prominently shaped development in the ASEAN countries as well as in China.

The opportunity to use low-cost labor in the PRC was particularly advantageous for businesses in Taiwan and Hong Kong, because for them transaction costs in the PRC were low. As Sung stresses in chapter 2, proximity, aided by common language and customs, made doing business on the mainland easy and cheap, once the mainland's economic system opened up.[5] Moreover, low transaction costs made it possible to initially move only the low-skilled labor-intensive stages of production onto the mainland, while retaining other activities in Hong Kong or Taiwan. Production chains were quickly created that crossed political boundaries and allowed Hong Kong and Taiwan to specialize in high-value services and technology-intensive production, while much of the ordinary manufacturing moved to the PRC. The combination created explosive growth on the Chinese littoral as production was moved onshore, permitting Hong Kong and Taiwan to sustain rapid growth at significantly higher income levels. The restructuring of existing export production networks into larger, lower-cost networks is the basis for the current division of labor.[6]

The sharp reduction in transaction costs among the China Circle participants reflects political changes specific to the Circle but was also part of a worldwide trend. Declining international transaction costs were the second of the fundamental changes in the economic environment that facilitated the emergence of the China Circle. As a result of declining costs, there is a global trend toward increased intra-industry trade, accompanied by increased intrafirm trade and associated with increased investment.

Alternately stated, there is a global tendency toward the geographic dispersion of production chains, and as a result an increasing share of international trade is made up of intermediate and capital goods.[7] As international transaction costs fall, it becomes profitable to locate different stages of the production process, characterized by different factor input needs, in different locations, according to where factor proportions and costs are most appropriate. Jean-François Huchet (chapter 8) calls this process "delocalization." The trend, which has been particularly prominent in the electronics industry, started with U.S. electronics firms in the 1970s. They were the first to significantly disperse their production facilities overseas, aided by favorable U.S. legal provisions; Southeast Asia, Taiwan, and Hong Kong were major beneficiaries.[8] The movement of certain labor-intensive stages of production chains to the China mainland is a specific case of this ongoing trend.

The third change in fundamental economic conditions was the collapse of raw material prices in the early to middle 1980s, and the sustained low raw material prices that have prevailed subsequently. These price shifts made the option of resource-based development unpalatable. It is sometimes forgotten that China's reintegration into the world economy at the end of the 1970s began with the rapid expansion of petroleum exports and that in 1985 petroleum and petroleum products still accounted for one-quarter of total Chinese exports ($6.7 billion).[9] Southeast Asian countries, particularly Malaysia and Indonesia, were also generating substantial portions of their exports in the mid-1980s from natural resources. In 1986–87 the ASEAN countries began a decisive turn toward a development strategy based on manufactured exports, at the same time that the PRC made its most important policy changes. As Tan shows, this policy shift in ASEAN countries led to substantial inflows of investment from Hong Kong and Taiwan (as well as Japan) and to an acceleration of economic growth. Simultaneous liberalization in China and ASEAN countries created a new competition for investment and third-country markets.

The speed and thoroughness with which restructuring has taken place in East Asia reflects the confluence of both regional and global factors. A catalytic role was played by currency realignments in 1985–87. These changes in relative costs, interacting with improvements in transportation and communication, tended to make firms act less as national "citizens" and more as international operators. Firm strategies became more important, since it became essential to be able to locate portions of the supply chain in the lowest-cost region, regardless of nationality. These changes in economic conditions and resultant firm strategies were the preconditions for the emergence of the China Circle.

The Emerging Division of Labor in the China Circle

There is already a significant literature describing and analyzing the growth of economic interactions among the Chinese regions.[10] This book builds on that literature to describe, analyze, and project the underlying economic dynamics that are propelling the region into the forefront of economic developments in the coming century. Over 70 percent of the direct investment in China came from Hong Kong and Taiwan. Trade data also show the changing relationships (figure 1-1). Total exports of the PRC, Hong Kong, and Taiwan grew from $85 billion in 1984 to $434 billion in 1995. During this period exports from the three Chinese regions to the rest of the world grew at a healthy 14 percent annual rate, increasing from $70 billion to $285 billion. But exports shipped among the three Chinese regions grew much more rapidly, at 23 percent a year, expanding from less than $15 billion to $150 billion.[11] As exporters increased sales outside the China Circle, they were also involved in much denser trade relations within the Circle.

Hong Kong played the central role in promoting these denser trade relations; a glance at Hong Kong data on re-exports reveals the main patterns. In 1995 Hong Kong's re-exports totaled $144 billion. The

Figure 1-1. *Exports of the China Circle, 1984–95*

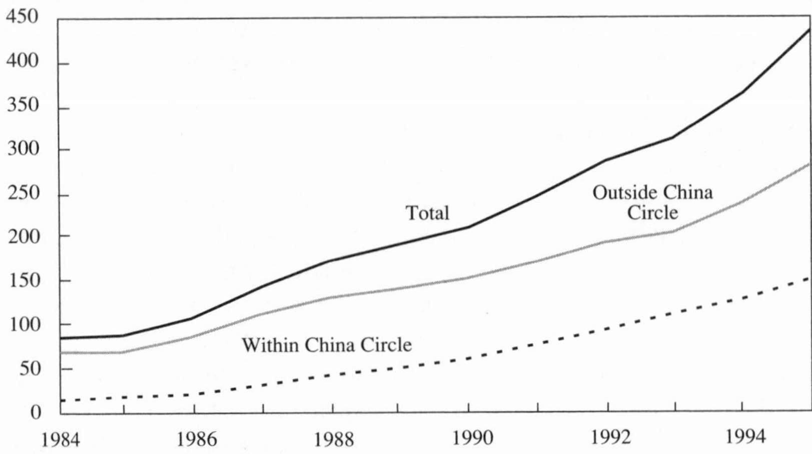

Source: International Monetary Fund, *Direction of Trade Statistics Yearbook 1996* (Washington, D.C.: International Monetary Fund, 1996), pp. 157–59, 240–42; Republic of China, Council for Economic Planning and Development, *Taiwan Statistical Data Book 1996,* Taipei: CEPD, 1996, p. 194. See also note 10 and related text, p. 11.

largest source country for re-exports was the PRC, which accounted for 57 percent of the total ($82 billion). Of PRC origin re-exports, 71 percent were consumer goods; three-quarters went to developed-country markets (34 percent to the United States alone). Taiwan was the third-largest source of re-exports (at $11 billion). Raw materials and semi-manufactures made up 73 percent (not including food and fuel), and capital goods were 18 percent. Thus 91 percent of Taiwanese re-exports were inputs into the production process, and 93 percent of these went to the PRC.[12] Taiwan's trade is thus an integral part of its investment on the mainland. More broadly, Hong Kong funnels equipment, components, and materials into China from Japan, Taiwan, and the United States and brings out consumer goods that it sells in developed-country markets.

As manufacturing production has moved to the China mainland, the southern coastal provinces have been industrializing rapidly, while Taiwan and Hong Kong are de-industrializing. The manufacturing labor force is declining in both Hong Kong and Taiwan. The Hong Kong industrial labor force declined from 0.93 million in 1985 to 0.386 million in 1995.[13] In Taiwan the manufacturing labor force reached a peak in 1987 at 2.8 million and declined to 2.4 million during 1995, a decline of 16 percent. Meanwhile, in the two provinces of Guangdong and Fujian, the industrial labor force increased from 6.1 million in 1985 to 10.75 million at the end of 1995.[14] Between them, Hong Kong and Taiwan have lost almost 1 million manufacturing jobs, while Guangdong and Fujian have gained more than 4.5 million. In fact, these data probably understate the total number of new manufacturing jobs in Guangdong and Fujian. There have been major flows of immigrants from other parts of China into these provinces, and some immigrants working in the informal sector are not captured in official employment statistics.

Hong Kong and Taiwan have both experienced substantial success in upgrading to higher-skilled activities, while simultaneously experiencing steadily rising incomes and relatively low unemployment. Hong Kong's restructuring has been especially thorough, because it has shed many industrial functions altogether and moved into greater specialization in services, particularly finance, transport, and telecommunications. In Taiwan, restructuring within the manufacturing sector itself has been the most impressive feature. Total manufacturing value-added has continued to grow even as manufacturing employment has dropped. Taiwan has moved into technologically more sophisticated products, while shedding low-technology products. Thus the upgrading of skills occurred in opposite and symmetrical ways in Hong Kong and Taiwan. Hong Kong moved smoothly out of manufacturing and into a variety of business services, such as finance, marketing, and accounting, yet it is currently seeking to upgrade its technological capabilities as well. Taiwan has been quite successful in improving technological capacities and moving into production and export of commodities at much higher technological levels, yet it seeks to also become a business operations and financial center.

A survey of foreign-invested firms in mainland China provides a breakdown of the sectoral composition of the accumulated stock of foreign investment as of the end of 1992. These data provide an excellent benchmark at the end of the first period of investment into the PRC, before the flood of incoming investment beginning in 1993, which was prompted by increased access to the mainland market. According to the 1992 data, Hong Kong was the largest investor in the mainland, and Taiwan was second, followed by Japan and the United States. When attention is restricted to investment in manufacturing, Hong Kong accounted for 64 percent of the total, and Taiwan 14 percent, reflecting the high concentration of Taiwan's investment in manufacturing. By this measure the United States was the third-largest investor, accounting for 6.1 percent of the total, and Japan dropped to fourth place, reflecting a higher proportion of investment in the service sector and the relatively cautious approach that Japanese investors took until after 1992.[15] Hong Kong and Taiwan had similar investment profiles: both were prominent in light and consumer manufacturing. Textiles and garments accounted for 26 percent of Hong Kong manufacturing investment, and 17 percent of Taiwanese. The electronics industry was second most important, accounting for 16 percent of Hong Kong manufacturing investment and 12 percent of Taiwanese manufacturing investment. Taiwanese investment has been somewhat more diversified than that of Hong Kong: plastic, rubber, and miscellaneous products are important, as are food, beverages, and wood products. Together with garments and electronics, these sectors account for fully 70 percent of Taiwanese manufacturing investment in the mainland. The sectoral patterns of investment reflect the fact that Hong Kong and Taiwanese investments, particularly through 1992, were dominated by export industries.

The sectoral composition of investments from Hong Kong and Taiwan, according to the same survey, differed significantly from those from the United States and Japan. U.S. manufacturing investment was prominent in chemicals, metal products, machinery, and transport equipment; Japanese manufacturing investment was most significant in metallurgy, machinery, and transport machinery. Through 1992 the strategies of U.S. and Japanese firms in China were directed toward import substitution investments destined for the Chinese market, or toward strategic investment designed to establish a presence for future diversified Chinese operations. As China has continued to reform its economy and open its markets, however, the distinction between investment from the China Circle and from the United States and Japan has probably begun to diminish. On the one hand, the orientation of Hong Kong and Taiwanese investments is changing. The phase of intensified reform beginning in early 1992, after Deng Xiaoping's southern tour, involved increased access to the Chinese domestic market for goods and services produced by foreign investors. Hong Kong and Taiwanese firms moved quickly to take advantage of these new opportunities, and investment from Hong Kong and Taiwan is increasingly motivated both by the search for low-cost export production sites and by the desire to establish access to the mainland market. On the other hand, U.S. and, especially, Japanese producers have become convinced—partly by the success of Hong Kong and Taiwan—that it is possible to use China as a low-cost production base and integrate China into global production networks. As Huchet shows, most Japanese electronics firms are now moving into northern China, and China is becoming an increasingly popular site for Japanese investment of all kinds. Thus the sectoral structure of investment in China from different sources may show some tendency to converge.

More recent figures show the output of foreign-invested firms in China in 1995, both in absolute value and as a share of total sectoral output (table 1-1).[16] Unfortunately, the sectoral data are for all foreign-invested enterprises (FIEs) and do not permit a breakdown by the nationality of the investor firm. Electronics is currently the single most important industrial sector for foreign investors in China. Electronics output from FIEs amounted to 14 percent of total FIE output. Textiles and garments are very important sectors, and the sum of textile and garment output was slightly greater than electronics output, although value-added in electronics was about 5 percent greater than that in textiles and garments together. Food and beverages were next most important, followed by transportation machinery, electrical equipment, and chemicals. The importance of transportation machinery and chemicals reflects the rapid growth of predominantly heavy-industry import substitution investment in the last few years. FIE output makes up a large share of output in several light manufactures, including sporting goods and plastics, and a surprisingly large share of transportation machinery and instruments.[17] FIEs accounted for a larger share of total output in the electronics sector than in any other sector, producing 60 percent of large-scale output.

A richer portrait of the division of labor emerges in chapter 6, by Chin Chung. Chung describes the overall process of industrial restructuring in Taiwan and the PRC, focusing on the migration of exporters. She also describes the development of the personal computer (PC) and components industry in Taiwan and shows that labor-intensive products enjoyed the largest cost advantage after production was relocated, and were in fact the first to be relocated. Taiwan affiliates began to produce keyboards and power supply units (the most labor-intensive prod-

Table 1-1. *Output of Foreign-Invested Firms in China, by Industrial Sector, 1995*[a]
Value in billions of Chinese *renminbi*

		Foreign invested firms	
Industrial sector	*Value of total output*	*Value*	*Percent of total*
Mining and timber	355.13	7.55	2.1
Food and beverages	519.59	119.47	23.0
Tobacco	100.42	0.56	0.6
Textiles	460.40	82.41	17.9
Garments	147.02	73.73	50.1
Leather	97.44	52.26	53.6
Lumber and furniture	63.16	18.22	28.8
Paper and printing	142.61	24.67	17.3
Sporting goods and stationery	37.11	18.60	50.1
Chemicals and refining	584.79	53.24	9.1
Pharmaceuticals	96.13	18.83	19.6
Synthetic fibers	80.99	11.14	13.7
Rubber products	61.99	15.52	25.0
Plastic products	112.77	37.63	33.4
Building materials	301.84	35.25	11.7
Ferrous metallurgy	366.02	23.00	6.3
Nonferrous metallurgy	137.23	17.34	12.6
Metal products	165.07	43.94	26.6
General purpose machinery	236.57	33.56	14.2
Industrial and specialized machinery	175.65	15.56	8.9
Transportation machinery	330.33	81.34	24.6
Electric equipment	259.43	63.14	24.3
Electronics	253.05	151.77	60.0
Instruments	42.57	16.88	39.7
Other	69.96	21.82	31.2
Manufacturing total	4,842.11	1,029.85	21.3
Utilities total	269.92	33.96	12.6
Industry total	5,494.69	1,071.40	19.5

Source: *Zhongguo tongji nianjian* (China statistical yearbook), *1996* (Beijing: Zhongguo Tongji, 1996), pp. 414–25.

a. Enterprises with independent accounting systems.

ucts) on the mainland in 1990, about two years after they began in ASEAN countries. In 1992 the production of monitors and motherboards began on the mainland, in this case with a one-year lag after ASEAN production. Chung also shows that China is accounting for a rapidly increasing share of Taiwan's offshore production of electronic components, the total of which is itself rapidly increasing.

The increased exchange of intermediate goods within the China Circle has not changed the fact that a large proportion of the markets for the region's final products lies in the OECD (Organization for Economic Cooperation and Development) countries, particularly the United States; the region as a whole is externally oriented despite its integrated production networks. The region continues to display a degree of market dependence on North America. One symptom of this has been the persistent U.S. trade deficit with the China Circle—formerly attributed primarily to Hong Kong and Taiwan, but now largely attributed to the PRC. For a series of specific commodities such as footwear, toys, and bicycles, a declining U.S. market share for Taiwan or Hong Kong has been matched by an increasing market share for the PRC.

Figure 1-2 shows the changing composition of the U.S. deficit with the China Circle. The United States began to run a large deficit with the China Circle in 1987, but at that time the deficit was predominantly with Taiwan. Between 1987 and 1996 the total U.S. deficit with the Circle grew from $26 billion to $47 billion. Far more striking than the growth in the overall deficit, though, has been the change in its distribution. Virtually all the deficit is now recorded against the PRC. The United States even runs a modest surplus with Hong Kong, in a dramatic turnaround from the 1987 situation. Clearly, this change in the distribution of the U.S. deficit is a direct result of the relocation of existing export production networks to the Chinese mainland. Total trade between the United States and the China Circle has grown more rapidly than the deficit, from $56 billion to $136 billion over the same period. As a result of this rapid growth, trade has become somewhat less unbalanced. U.S. exports to the China Circle were only 37 percent of U.S. imports in 1987 but grew to 49 percent of U.S. imports in 1996.

Government Policy and Politics

Trade and investment links increased throughout the East Asian region in the 1980s, usually without active government involvement and without explicit government-to-government agreements. There was very little explicit government facilitation of the process before the

Figure 1-2. *U.S. Trade Deficit with the China Circle, 1987–96*

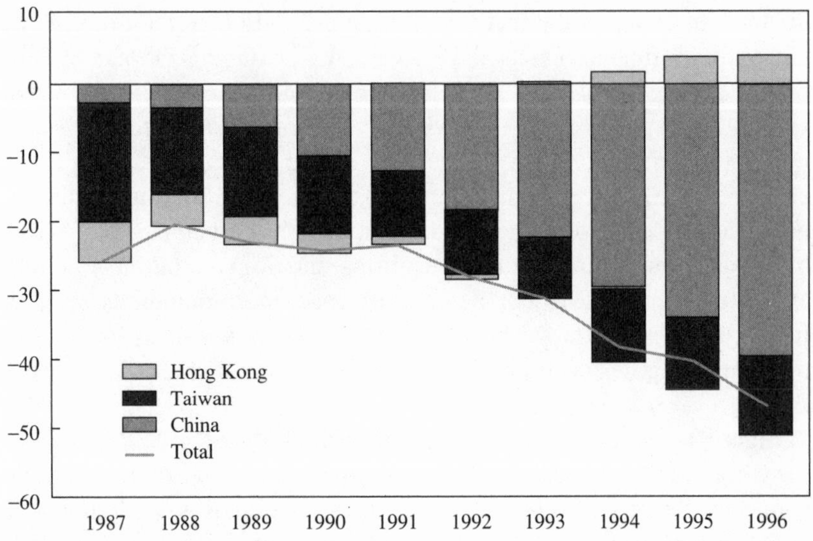

Source: U.S. Department of Commerce, Office of Trade and Economic Analysis, as of March 1996 at http://www.ita.doc.gov/industry/otea/usfth/hili.html.

formation of Asia-Pacific Economic Cooperation in 1989. Only with the acceleration of the APEC process, usually dated to the Seattle APEC summit in November 1993, did governments start to take an active role in concerted trade and investment policies. The China Circle governments are extreme examples of this general absence of a multilateral trade policy. The only significant formal bilateral agreement was the 1984 agreement on the future of Hong Kong, signed between China and the United Kingdom! In any case, the Hong Kong agreement, though of tremendous importance because of its wide-ranging political impact, had almost no direct effect on the developing economic ties between Hong Kong and the PRC.

Instead, relations among China, Taiwan, and Hong Kong generally correspond to the pattern of competitive unilateralism that Tan discerns in ASEAN-China relations. Governments recognize and respond to policy changes in competitor nations, as well as to changes in regions that are potential sources of investment. Governments do not appear to be actively shaping development as much as they do to be responding to changing opportunities. This largely more passive government role corresponds to the desire of governments to facilitate incoming direct investment. Whereas China also attempts to operate an industrial policy that it hopes will shape future development, most of the China Circle development described here is the result of a general accommodation of investment rather than of specific government initiatives. Of course, policy reform was essential to accommodate the growth and investment process. What was surprising was the relatively quick and effective way in which policy in the PRC responded to changing external conditions (see chapter 3).

The economic cooperation between the People's Republic of China and Taiwan has developed within the context of continuing rivalry and competition between two political and economic systems. Both sides have tried to use economic cooperation as a carrot (and sometimes a stick) in the struggle to prevail in the international arena and in the political competition more generally. With the growth of economic cooperation, the rivalry between the PRC and Taiwan has shifted from one played out at a wary distance to one fought out at close quarters and with more subtlety. The rivalry is, however, no less intense for that, nor are the passions necessarily muted. In 1995 and 1996 the PRC lobbed missiles near Taiwan's busiest harbors, demonstratively asserting its "right" to use force in its quest for political reunification. Nevertheless, trade and investment across the Taiwan Strait continued to increase despite a sharp drop in trade and investment during the early months of 1996 when tension was at its height. In the meantime Hong Kong has prepared for the return to PRC sovereignty on July 1, 1997, with a curious political configuration in place: the wealthiest businessmen are supporting a nondemocratic government imposed by the communist government in Beijing in disregard of the expressed wishes of the majority of the people of Hong Kong. Politics and economics are intertwined everywhere in the China Circle, but rarely does the relationship between the two correspond to a simple model in which policy instruments are manipulated to lead to economic outcomes.

Firm Networks and Industry Strategies

The economies of Hong Kong and Taiwan are dominated by small firms. In contrast to both Japan and Korea, in which large firms and conglomerates play a dominant role in the economy, in both Hong Kong and Taiwan small firms account for the bulk of manufacturing employment and output and a majority of manufactured exports.[18] Large firms now play a significant role in both economies, but they have emerged fairly recently from the competitive ferment of numerous small firms, and their relative importance is still much less than in Japan or Korea. The predominantly small-firm environment has tended to produce both characteristic forms of integration into the world economy and distinctive patterns of industrial strategy. At the same time, in the transitional economy of mainland China, small and medium-size firms have clearly emerged as the most dynamic sector in the economy. Under the planned economy, China was dominated by large firms, but as the economy has "grown out of the plan," the vibrant small-firm sector has come to play a larger and larger role.[19] The size distribution of firms in China has changed in the direction of that of Hong Kong and Taiwan, and small and medium-size firms have also played the predominant role in the emergence of the China Circle within the PRC.

Small China Circle firms tend to form fluid, "open" networks of transactions and contracting and subcontracting. Existing studies show that these small firms have high birth and death rates, that employees are constantly leaving existing firms to start up new ones, and that firms are linked by dense networks of transactions, contracts, and personal ties.[20] Despite their small size, firms are deeply engaged in international transactions. In both Hong Kong and Taiwan, small firms have learned to export and engage in other sophisticated international transactions. Often those firms have been brought into international networks by buyer firms from developed countries that serve as markets, frequently from Japan and the United States. A striking aspect of the export success of both Taiwan and Hong Kong is that it has been significantly based on order-taking from importers in developed-country markets (especially the United States). Again, the contrast is with Japan—and to a lesser extent with South Korea—where export networks have been controlled by large corporations within the exporting country, with international trading companies (*soga sosha* in Japan) playing a prominent role. Rather than independently developing new products, as a Japanese firm might do, exporters in Taiwan and Hong Kong have typically responded to requests for new products from a U.S. importer. These firms have become exceptionally skilled at producing reliably and quickly to a set of specifications provided by an importer.[21] This ability in turn has given the China Circle exporters an enormous advantage in any industry in which speed, or time to market, and the ability to provide precisely the attributes the market demands are the critical determinants of success. That is true both in sectors in which the need for timeliness is the result of fashion, such as clothing, shoes, and even toys, and in sectors driven by rapid technological change, such as electronics.

The export networks of China Circle producers are open to importers, who frequently travel to these regions to procure commodities. This results in an important difference between China Circle international production networks and those operated by Japanese firms. Japanese import trade tends to be dominated by intrafirm trade; much of the recent rapid growth in Japanese imports has come from overseas subsidiaries of Japanese firms and branches of Japanese trading firms.[22] No such generalization is possible for China Circle trade; a bewildering variety of transactional forms characterizes trade both among the China Circle economies and with external partners. The diversity and general openness of China Circle trading networks provide yet another perspective on the importance of Hong Kong, as described by Sung (chapter 2). Hong Kong has to be an efficient order-taker. Buyers must be able to travel freely to Hong Kong, stay in constant communication with home country corporations, and yet network quickly and thoroughly with local producers. Hong Kong's essential role as intermediary is related partly to its superb infrastructure facilities and partly to its unique ability to provide an interface between flexible networks of small-scale producers and demanding international markets. What is most striking is that Hong Kong's businesses have been able to extend those networks into Guangdong province and even into hinterland China, while still maintaining crucial speed and responsiveness to the market.

The predominance of small firms in both Hong Kong and Taiwan has also shaped their characteristic approach to developing individual industries. Here a simple contrast between a South Korea model and a Taiwan model might be useful. Korean businesses, particularly the large, diversified industrial conglomerates (*chaebol*), have frequently chosen to enter industries in which large initial investments are required in order to establish large factories. That has often included entry into activities unrelated to existing lines of business. With access to abundant capital, the *chaebol* have been able to reach minimum economic scale quickly, even in sectors characterized by economies of scale through very large plants. Once these plants have been established, Korean firms have relied on a mastery of process technologies, combined with a low-cost labor advantage, to enter new markets. By contrast, Taiwanese businesses have typically begun as small-scale start-up businesses, emerging through technology transfer from foreign firms or through the imitation of existing businesses. Employees have commonly left an existing firm and set up a new firm, which may begin supplying a component or assembly to the original firm. Subsequently, the most successful new firms may grow to be large firms, but as a rule they do so initially on the basis of competitive success in a single line of business. Clearly, this has tended to produce an economy that is less concentrated in large scale-intensive plants than Korea is, but that has a greater ability to produce diversified products in smaller batches in response to market needs. Both Taiwan and Korea have moved on a similar path from labor-intensive, low-technology manufactured exports to progressively more technology- and capital-intensive products. Different firm structure in the two economies has, however, tended to produce contrasting trajectories in this otherwise similar developmental transformation.

The Electronics Industry

The choice of electronics for a sectoral case study is a natural one. The industry is large and also strategically important to businesses and consumers in the United States.[23] Moreover, electronics is a natural choice to bring considerations of technology, firm strategy, and world market conditions into the analysis. Electronics is an important industry in both Taiwan and Hong Kong and has recently been transferred to the PRC at a rapid pace. When analyzing the electronics industry, one tends to focus on Taiwan, because the growth of Taiwan's computer industry over the past decade has been particularly impressive. Taiwan now has the strongest capability in the electronics industry of the China Circle economies. Chung (chapter 6) provides a particularly clear and detailed description of the dramatic growth of Taiwan's computer industry, including both the growth of technological capabilities and the process of restructuring that has moved substantial capacity to the mainland.

The general observations on firm size in Hong Kong and Taiwan apply fully to the electronics industry. In Hong Kong the number of electronic component suppliers increased from 9 in 1966 to 1,016 in 1987; and there were 454 firms assembling electronic watches in 1984. Firm birth and death rates are high, and extensive subcontracting relationships exist.[24] Taiwan today is home to more than 4,000 firms that produce a broad mix of PC-related products and electronic components. Almost all these companies started out as small companies, often with just a handful of employees. Chung and Dieter Ernst (chapter 7) clearly show how the dynamism of the industry is related to its small-scale entrepreneurial background. Taiwan's flexible producers, with selective government technical assistance, have learned how to be fast followers, only a half step behind the technological frontier.[25]

The contrast between the Taiwan model and the Korean model in electronics is exemplified by the development of the integrated circuit (IC) industry in the two economies. In Korea the major *chaebol* pushed into the production of dynamic random-access memory (DRAM) chips on a massive scale in the 1980s and 1990s and, given their mastery of process technologies (placing them near the best-practice industry leaders) and low costs, quickly emerged as competitors on a world scale, particularly to the Japanese industry. Taiwan, by contrast, was an insignificant producer of DRAM chips until quite recently. Instead, the Taiwanese IC industry developed first through concentration on application-specific integrated circuits (ASICs), which must be custom designed for use in specific applications (often consumer goods) and which are typically produced in much smaller batches than DRAM chips. Only since 1995 have large wafer fabrication facilities (fabs) for DRAM chips been built in Taiwan. Competitiveness in DRAM chip production requires cheap financing and mastery of process technology, while competitiveness in ASIC production requires strong design capabilities and efficient adaptation to market demands. Because DRAM chip production requires large capital investments (with significant lead times) relative to current costs, the industry tends to cycle through periods of surplus and shortage. During periods of shortage, the industry is extremely profitable; during periods of surplus, losses can be staggering. Korea's industry has experienced both extremes over the past several years,[26] whereas Taiwan's more diversified and flexible industry is less buffeted by market fluctuations.

Michael Borrus (chapter 5) describes how the electronics industry, especially the computer industry, has been evolving toward a technological foundation of networkable, open systems. In this environment, standards are not proprietary, and any producer can supply the market if it can provide the right combination of speed, quality, and connectivity. This environment strongly favors the China Circle producers (as well as those based in Singapore), who already possess competitive strength in precisely these areas. At the same time the technological shift has worked to the advantage of the United States—which predominantly establishes the standards to which the industry works—and has inevitably reduced the comparative competitive advantage of Japanese producers, who tend to be committed to a technological strategy and organization of production less adaptable to open standard competition. Combined with the dramatic drop in costs occasioned by the transfer of labor-intensive production to the PRC, these technological changes have greatly increased the competitiveness of the China Circle electronics industry. Indeed, there is a general match between the capabilities of the United States and the China Circle, in that one is strongest in fundamental research, product design, and the creation of standards, while the other is strongest in manufacturing technology and industrial market responsiveness. There is also a more fortuitous match between the United States and the China Circle, in that the system of open technological standards that emerged out of tumultuous competition in the United States provides an excellent fit with the system of open firm networks that emerged out of tumultuous competition in the China Circle economies. There are thus surprisingly broad and deep economic links between the United States and the China Circle.

Despite the impressive achievements and linkages, in an important sense the China Circle remains dependent for supplies on Japan. As Ernst and Borrus stress, Japan provides much of the capital equipment on which industrial production is based, and it also continues to dominate the production of certain high-tech core components. Japan acts as supplier both through direct sales of goods and equipment embodying high technology and through investment in key sectors. This supply dependence is manifest in a broad range of productive sectors. In electronics it is evident in the continuing Japanese dominance, at least within Asia, of the production of many core components. While most central processing units (CPUs) come from the United States, flat-panel displays come predominantly from Japan, as do the largest memory chips. Indeed, Taiwan has a persistent trade deficit with Japan that largely reflects Taiwan's dependence on Japan for key components as well as for many types of industrial equipment. These characteristics serve as a reminder that the China Circle is open in many respects to actors outside the Circle, since it has important links both to high-tech suppliers and to third-country markets for final goods, especially consumer goods.

The Electronics Industry in the People's Republic of China

The most impressive result of the China Circle interactions within the PRC electronics sector has been the creation of a large and diversified, though relatively low-tech, electronics supply base in Guangdong and Fujian provinces. The center of gravity of China's electronics industry has shifted south to those provinces, that is, into the China Circle. In 1985 Guangdong and Fujian accounted for only 12 percent of electronics value-added in all China; by 1994 their share had jumped to 31 percent.[27] Guangdong produces more electronics output than any other province and leads by substantial margins in production of both consumer electronics and computer-related products. Electronics exports have increased rapidly, reaching $12.4 billion in 1995. Moreover, as noted earlier, foreign investment is more prominent in electronics than in any other PRC industry. That is true in exports as well as production: in 1995 FIEs accounted for 57 percent of total electronics exports.[28] Judging by U.S. import statistics, these exports consisted overwhelmingly of consumer electronics plus in recent years significant quantities of computer peripherals and subassemblies.

The large role of foreign-invested enterprises in the Chinese industry reflects the dynamism of the world electronics industry and the potential importance of China both as a low-cost production site and as a large and growing market. But it also reflects the stark failure of the Chinese government's attempt to foster a competitive electronics industry based on indigenous technological resources and predominantly on state-owned enterprises. Huchet (chapter 8) discusses the evolution of policy and the main reasons for its failure. In brief, the Chinese government has tried repeatedly to build up "national champions": large, capital-intensive enterprises that use advanced technologies and are able to achieve economies of scale. Repeatedly, this effort has failed. Failure has been most notable in the effort to create an integrated circuit industry. As Huchet describes, only one state-owned firm has managed to stay in the business at anything near economic scale, and only after many millions of dollars of investment. Output projections have been repeatedly scaled back. Using IC line width as an indicator of technology level, Shanghai IC producers—then among the best in China—who were visited in 1985 reported using 5 micron lines in current production, placing them about five years behind Taiwan and more than ten years behind the world leaders (see Chung, table 6-13). In mid-1995, ten years later, after large investments and expensive purchases of foreign technology, China's most advanced producers were struggling to get 1 micron and 0.8 micron line width IC production lines in operation and producing smoothly. That is, they were still five years behind Taiwan and about nine years behind the world leaders, thus almost completely failing to shorten the technological lag in a key process technology. In accessing international technology, Chinese producers had taken so long to get contracts signed, equipment installed, and production ramped up that by the time processes were working smoothly, they were already obsolete, and a new round of international technology purchase was required.[29]

Huchet describes the development of Chinese policy toward electronics, as well as its relation to Chinese technological capabilities and to shortcomings of the Chinese economic system. Indeed, even Chinese industrial policymakers have begun to respond to this tremendous differential in performance. Particularly within the Ministry of Electronics, development strategy has shifted from its traditional emphasis on creation of high-technology prototypes and instead adapted to stress production of personal computers, components, and peripherals. Long-term objectives have increasingly been formulated in terms of an information policy for society as a whole. The immediate effect on the centralized drafting of industrial policies has been paralysis and confusion. A national meeting was convened in January 1995 to draft an industrial policy document for the electronics sector. Since that time, however, disagreement about principles and conflicts over bureaucratic turf and funding have blocked agreement, and as of early 1997 no industrial policy document has emerged. China has maintained a direct state role in promoting IC development, but de facto industrial policy for the electronics sector as a whole has been scaled back and adapted to the real opportunities that China faces.[30] The failure to come up with a recent industrial policy has in fact been almost wholly beneficial, since it has allowed decentralized decisionmaking to continue, much to the benefit of electronics development in China.

Government industrial policy had been focused on the Shanghai and Beijing areas: the decline of centralized industrial policy paved the way for the emergence of Guangdong as the center and most dynamic region of the electronics industry. The growth of a supply base in Guangdong is related to the existing capabilities both of Hong Kong and Taiwan firms and of domestic PRC capabilities. The electronics industries in both Hong Kong and Taiwan got their start with consumer electronics, but the two subsequently followed different trajectories. Taiwan moved steadily into industrial electronics, and especially computers, while Hong Kong generally remained specialized in consumer electronics. These different paths are reflected in the pattern of industries that were transplanted to Guangdong during the 1980s and 1990s, and indirectly reflected in the pattern of exports from the PRC. Hong Kong firms were the first to shift production to China, and as a direct result, by 1994 China was able to export $6.7 billion worth of telecommunications and sound equipment. The most important products were telephone handsets and tape players assembled in China from imported components. Since 1994 growth of these products has moderated, and $8 billion of such equipment was exported in the first eleven months of 1996. As Chung shows, the restructuring of Taiwan computer production was just beginning in the early 1990s, and U.S. computer firms followed the Taiwan example by a few years. As a result, Chinese exports of data processing equipment and components, which were $2.7 billion in 1994, have grown exceptionally rapidly since, amounting to $5.9 billion in the first eleven months of 1996.[31]

Domestic PRC capabilities were not initially present in the electronics industry in Guangdong. Instead, the liberal trade and investment regime created in the China Circle (discussed in chapter 3) tended to divert trade and output growth from other regions of China to the provinces of Guangdong and Fujian, and especially to the local special economic zones (SEZs).[32] This process is illustrated by China's two largest computer companies, Lianxiang (Legend) and Great Wall. Both these companies had their origins in Beijing in the mid-1980s. Legend was a spin-off from the Chinese Academy of Sciences; Great Wall was set up within the Ministry of Electronics as part of its conversion from military to predominantly civilian orientation. Their birth from the fertile soil of Beijing's research institutes reflected Beijing's abundant human resource base, as well as the entrepreneurial climate that fostered the development of a commercial high-tech district (*zhongguancun*) in northwest Beijing, near the main universities and national research institutes.[33] Both companies maintain a Beijing presence today, but both found that in order to grow, they needed to transfer substantial parts of their trade, production, and research to Guangdong. In the economic environment of Beijing, it was too difficult for these firms to access components and technology from the world market, and thus too difficult for them to find an appropriate niche within the rapidly evolving international division of labor.

Legend established a number of subsidiaries and joint ventures in Hong Kong and the Shenzhen SEZ during the 1980s. These connections allowed the company to concentrate on the production of motherboards and video cards, in which, by 1995, it had achieved world market shares of 3 percent and 10 percent, respectively. Clearly, these successes depend on easy access to, and integration with, the international market, since the key components are imported, mounted on boards, and re-exported. Indeed, the company's philosophy explicitly calls for the imitation of proven technologies and concentration on middle stages of production, with both supply and sales on the world market (*liangtuo Zaiwai*). The company also produces PC clones under its own brand for the Chinese domestic market, primarily assembling imported components; in 1995 it achieved a 10 percent market share, ranking the company third.[34] Great Wall, after an uneven beginning, has followed a similar strategy. A recent linkup with IBM (which will purchase $1 billion worth of computer components and peripherals in China in 1996, according to Chinese sources) provides Great Wall both with a source of technology and components for domestic sales and with a customer for its power supplies, boards and cards, and other peripherals.[35]

One way to describe the evolution of China's electronics industry would be to say that the government tried but failed to develop an industry along the Korean model—one based on large-scale investment in capital-intensive plants the success of which depends on mastering sophisticated process technologies. Simultaneously, entrepreneurial individuals, working under the aegis of nominally state-owned corporations, succeeded in developing industry segments in a pattern much like the Taiwan model.[36] To do so, they had to relocate to Guangdong or Fujian, where they could rely on a greater degree of economic liberalization and proximity to Hong Kong. Crucial in this process was their ability to link up with firms in Hong Kong and Taiwan and benefit from incoming investment from those sources. The result of this process has been the gradual creation of a supply base in that area. The region is now a major producer of consumer electronics, predominantly the assembly of medium- and lower-quality final goods, but including some component production, such as simple LCDs (liquid crystal displays). At the same time it has developed substantial production capacity in a wide range of computer products. In 1996 the PRC produced 12 million motherboards, about 30 percent of global output, and 8 million monitors, about one-eighth of the total number produced worldwide. Most of this output comes from affiliates of Taiwanese firms. A process of gradual upgrading seems to be under way. IBM recently decided to build a plant in the Shenzhen special economic zone to produce head gimbals, a component for hard drives. The assembly of this component is very labor intensive, but the particular type to be assembled (magnetoresistive) also has fairly demanding technical requirements.[37]

The Chinese market for personal computers has also been growing rapidly. From a mere 100,000 units shipped in 1991, volume surged to a total estimated between 1.66 million and 2.1 million units in 1996. American brands dominated this market, with IBM selling the most units, and Compaq the top money-earner. The Chinese Legend brand surged to second place in sales, while Great Wall was seventh.[38] In fact, there is considerable ambiguity about the identity of the various PC brands in China. The U.S. brands are predominantly manufactured in the China Circle, usually in Taiwan. Conversely, many "manufactured in China" machines are simply assembled from imported components, in order to take advantage of the lower tariff rates on components than on finished machines. These machines could carry either a U.S. or Chinese brand name, depending on the status of the assembly operation. The ambiguity reflects the relatively open character of the industry and the hybrid ownership forms that are increasingly important. This supply base is now substantial enough to occupy an important niche in the Asian division of labor in the electronics industry.

One modification is required to this otherwise simple story. In the telecommunications market, rapid growth and continued government control have produced a very different environment from that in the electronics market. China has been investing over 1 percent of GDP in telecommunications annually since the early 1990s, and the result has been a dramatic growth in the number of telephones, from 10 million during 1989 to 65 million in October 1996, for a penetration rate of 5.8 percent. At the same time there has been dramatic upgrading of telecom-switching technology and extensive laying of fiber optic cable.[39] As a result, the Chinese government has been able to attract major investments from large multinational corporations not only interested in supplying the Chinese telecom network but also willing to produce in China in order to do so. Simply put, the Chinese government has much more leverage in this (large) market segment, since it is the dominant customer. The largest foreign-invested electronics firm in China is not in the China Circle at all. The largest firm is Motorola (in Tianjin), with just short of $886 million (7.4 billion yuan) in sales in 1995 and virtually no exports.[40] These multinational corporations are predominantly from the industrialized countries—beginning in about 1995 they began to build large factories that will use submicron technology in IC production. However, almost all output will go into the telecommunications equipment that these firms will produce and sell to the telecom network. In general, these ventures are beyond the capability of China Circle firms. As both Ernst and Huchet stress, this is a field of endeavor in which China attracts developed-country investment on terms rather different from those that prevail in the China Circle.

Links to the World Electronics Industry

It is clear from the preceding discussion that the electronics industry in the China Circle is not in any sense developing in isolation from the world electronics industry. This is true both in the general sense that the electronics industry in Asia grew up through transplant and imitation rather than through creation, and more specifically in that the rapid growth of electronics production in the China Circle in recent years is the result of a process of restructuring and upgrading of existing networks rather than of indigenous PRC development or even PRC imitation. Thus the growth of the electronics industry in the China Circle is not by any stretch of the imagination an autonomous process, and it cannot reasonably be analyzed in the context of the China Circle alone. The market leaders are in the United States and Japan, and the rapid growth of China Circle production reflects the response of market leaders—as well as "fast followers" like Taiwan—in seeking to control costs. Yet those market leaders have responded to restructuring pressures in different ways, and these different response strategies have different impacts on the China Circle. Because the Circle is a region in which capabilities are increasing rapidly as a result of restructuring pressures, the specific form that restructuring takes matters a great deal.

In the second part of the book, four authors deal with the relationship between the China Circle and the global electronics industries from four different perspectives. Borrus (chapter 5) views it, in a sense, primarily from an American perspective; he stresses the increase in joint U.S.-China Circle competitiveness with Japan from the late 1980s and the role of U.S. investment and U.S.-China Circle technical collaboration in the creation of a regional supply base in the China Circle. He sees this process as being driven by the convergence between the U.S. ability to increasingly set the standards that prevail in open, networkable systems and the manufacturing capabilities of the China Circle, which are ultimately based on open, networked systems of small businesses. Chung (chapter 6) views the situation primarily from a Taiwanese perspective; she stresses the steady increase in Taiwan's capabilities and the pressures on Taiwan to restructure the electronics industry created by appreciation and increasing labor costs. Although she recognizes the existing close relationship between Taiwanese and U.S. electronics firms, she sees Taiwanese firms as being on the verge of growing out of a technologically subordinate role to the U.S. firms in a way that might lead to an increasingly competitive relationship. Ernst (chapter 7) views the growth of the China Circle primarily from a Japanese perspective; he stresses the closed nature of Japanese production networks in the late 1980s and the pressures on Japanese firms to open up those networks in the late 1990s. The China Circle producers in this view play the role both of competitors to the Japanese firms and increasingly of attractive partners and collaborators. Thus Ernst implicitly envisages a gradual intertwining of firm networks from Japan with those from the Circle. Finally, Huchet (chapter 8) views China Circle electronics primarily from a mainland Chinese perspective, stressing the failure of Chinese industrial policy to create an autonomous Chinese electronics industry and the continuing limitations on the PRC's ability to absorb technology via foreign direct investment.

All four authors share an interest in the interaction between corporate strategies and the gradual accumulation of capabilities in the host countries. From their different perspectives, and with their different emphases, they substantially agree that a powerful, long-term process of technology transfer and accumulation of capabilities is under way in the China Circle. Both Borrus and Chung, in stressing the impressive rise of the Taiwanese computer manufacturing industry, seem to reflect an attitude of long-term historical optimism that is definitely in the air in many of the China Circle regions. One often hears there a confidence that the region itself, with its rapid growth and technological dynamism, represents the future. Ernst and Huchet provide a cautionary note. Despite the dynamism of the China Circle, it is still a long way behind the market leaders in Japan and the United States. More crucially, its long-term fate depends on the ability of the PRC to continue and intensify a process of social, economic, and political reform that has only begun to transform that huge society. The authors of this volume hope that their concentration on the dynamics of the economies and of a single industry can serve as a lens through which the broader and less tangible questions about the future of the China Circle region can be brought into sharp focus.

Notes

1. Robert Scalapino, "The United States and Asia: Future Prospects," *Foreign Affairs*, vol. 10 (Winter 1991–92), pp. 19–40.

2. International Monetary Fund, *Direction of Trade Statistics Yearbook, 1992* (Washington, 1992), p. 2; IMF, *Direction of Trade Statistics Quarterly* (September 1996), p. 2; China State Statistical Bureau, *Quanguo gesheng zizhichu zhixiashi lishi tongji ziliao huibian (1949–1989)* (Compilation of provincial historical statistics, 1949–1989) (Beijing: Zhongguo Tongji, 1990); and China General Administration of Customs, *China Customs Statistics, 1995*, vol. 12, pp. 12–15.

3. Barry Naughton, "China's Emergence and Prospects as a Trading Nation," *Brookings Papers on Economic Activity, 2:1996*, pp. 312–13.

4. Unfortunately, there are no good data on interprovincial commodity flows within the PRC. Hong Kong actually re-exports significant amounts of goods from one Chinese province to another: $4.7 billion in 1995, according to Census and Statistics Department, *Annual Review of Hong Kong External Trade, 1995* (Hong Kong, 1995), p. 88. The dense network of commodity flows, combined with continuing restrictions on factor movements, is characteristic of other cross-border production regions, such as the San Diego-Tijuana area and the Singapore "growth triangle."

5. Indeed, the importance of low transactions costs can be demonstrated by the fact that both U.S. and Japanese businesses frequently choose to manage their interests in

the People's Republic via intermediaries in Hong Kong and Taiwan. For example, the U.S. athletic shoe producer Nike subcontracts its mainland production to Taiwan intermediaries, while the Japanese firms TDK and Yamaha both operate in the mainland through Taiwan subsidiaries. See China Economic News Service, "Japan Firms Use Taiwan as Base to Cultivate Mainland Market," February 27, 1993, via Reuters Textline. Hong Kong cases are even more numerous; see Huchet (chapter 8) for additional examples.

6. Beginning in 1992 this basic dynamic was augmented as the Chinese government began to offer investors expanded access to the Chinese market. The flow of investment became a flood in 1993–94, and investors increasingly followed mixed strategies in which sales to the Chinese domestic market played an important role.

7. Richard Grant, Maria C. Papadakis, and J. D. Richardson, "Global Trade Flows: Old Structures, New Issues, Empirical Evidence," in C. Fred Bergsten and Marcus Noland, eds., *Pacific Dynamism and the International Economic System* (Washington: Institute for International Economics, 1993), pp.17–63.

8. A. J. Scott, "The Semiconductor Industry in South-East Asia: Organization, Location and the International Division of Labour," *Regional Studies,* vol. 21, no. 2 (1987), pp. 143–60.

9. *Zhongguo tongji nianjian* (China statistical yearbook), *1986* (Beijing: Zhongguo Tongji, 1986), pp. 569–70. Hereafter *China Statistical Yearbook.*

10. The most important recent works include Robert Ash and Y. Y. Kueh, "Economic Integration within Greater China: Trade and Investment Flows between China, Hong Kong, and Taiwan," *China Quarterly,* no. 136 (December 1993), pp. 711–45; Chang Jung-feng, *T'ai-hai liang-an ching-mao kuan-hsi* (Economic and trade relations across the Taiwan Straits) (Taipei: Kuo-chia Cheng-ts'e Yan-chiu Tsz-liao Chung-hsin [Institute for National Policy Research], 1989); Randall Jones, *The Chinese Economic Area: Economic Integration without a Free Trade Agreement* (Paris: OECD Department of Economics and Statistics, 1992); Kao Charng (principal investigator, Chung-hua Institute of Economic Research), *Taishang yu waishang zai dalu touzi jingyan zhi diaocha yanjiu—yi zhizaoye weilie* (Survey research of the experience of Taiwan and foreign investors on the mainland—the example of manufacturing) (Taipei: Ministry of Economics, April 1994)(hereafter *Survey Research*); Kao Hsi-chun, Li Ch'eng, and Lin Tsu-chia, *T'ai-wan t'u-p'o: liang-an ching-mao chui-tsung* (Taiwan breakthrough: tracking cross-Strait trade and investment) (Taipei: T'ien-hsia Wen-hua, 1992); Sumner La Croix, Michael Plummer, and Keun Lee, eds., *Emerging Patterns of East Asian Investment in China: From Korea, Taiwan, and Hong Kong* (Armonk, N.Y.: M. E. Sharpe, 1995); Qi Luo and Christopher Howe, "Direct Investment and Economic Integration in the Asia Pacific: The Case of Taiwanese Investment in Xiamen," *China Quarterly,* no. 136 (December 1993), pp. 746–69; Thomas P. Lyons and Victor Nee, *The Economic Transformation of South China: Reform and Development in the Post-Mao Era* (Ithaca: Cornell University East Asia Program, 1994); and Nobuo Maruyama, *Kanan keizaiken* (The South China economic region) (Tokyo: Ajia Keizai Kenkyujo, 1992).

11. These data include re-exports as well as trade in both intermediate and final goods. International Monetary Fund, *Direction of Trade Statistics Yearbook, 1996* (Washington, 1996), pp. 157–59, 240–42; and Council for Economic Planning and

Development, *Taiwan Statistical Data Book, 1996* (Taipei, 1996), p. 194. Important gaps in the data remain. Sung's analysis throws light on a broader data problem that has not been entirely resolved. Problems are especially severe for Taiwan's trade data, since certain types of trade with the mainland have been forbidden, at least until recently. Taiwanese figures substantially understate both exports to and imports from the mainland, which are sometimes imported under false country-of-origin invoices. The Hong Kong statistics can partially resolve questions about mainland trade but not eliminate these issues. Paradoxically, recognition by the Chinese authorities that its statistics were not capturing the ultimate destination of most of its exports caused them to shift the principles of export reporting beginning in 1993. But because the changeover was not complete, the short-term result has been that the data have become less informative; it is now more difficult to correct the data by calibrating it with Hong Kong data. In this calculation, exports from China to Hong Kong have been approximated by using the growth rate of imports to Hong Kong from China, as reported by Hong Kong authorities.

12. Japan and the United States were the second and fourth largest sources of re-exports, respectively, with $17 billion and $7 billion. The destination and composition of commodities by end use for Japan and the United States were virtually identical. For both countries, 70 percent of re-exports went to the PRC, 33 percent of the total were capital goods, 41–44 percent raw materials and semimanufactures, and 20 percent consumer goods. Census and Statistics Department, *Annual Review of Hong Kong External Trade, 1995* (Hong Kong), pp. 78–79, 88–89. In chapter 2 Yun-wing Sung makes clear that re-exports are only one part of the extensive trade channeled through Hong Kong. But there is no compelling reason to expect the commodity composition of other types of trade to differ substantially from re-export trade.

13. Census and Statistics Department, Hong Kong, as reported in *Hong Kong Industrialist*, March 1996, p. 95.

14. Council for Economic Planning and Development, *Taiwan Statistical Data Book, 1996* (Taipei, 1996), p. 19; and *China Statistical Yearbook, 1986*, p. 125; *1996*, p. 92. China data include Hainan with Guangdong and include mining and utilities with manufacturing; in both cases, for comparability with earlier figures.

15. The data were gathered through a collaborative project between the Chung-hua Institute in Taipei and the State Statistical Bureau in Beijing and show cumulative realized investment 1979–92 (see table 6-2). The data are from Kao Charng, *Survey Research.*

16. The data reported in table 1-2 cover only enterprises with independent accounting systems, classified at the township level and above. It thus *excludes* village-level collective enterprises and virtually all private firms. Total output from independent accounting enterprises accounts for 60 percent of total industrial output, but 83 percent of state-owned enterprise output and 98 percent of foreign-invested firm output. The category corresponds reasonably well to the large-scale industrial sector, but there may be important omissions of coverage. *China Statistical Yearbook, 1996*, pp. 414–15, 424–25.

17. The large foreign share of garment industry output is misleading: most garment factories in the PRC are village-level collectives or private firms, data for which are not included in table 1-2.

18. Victor Sit and Siu-lun Wong, *Small and Medium Industries in an Export-Oriented Economy: The Case of Hong Kong* (University of Hong Kong, Center of Asian Studies, 1989). In Taiwan in 1993, small and medium-size enterprises accounted for 69 percent of the total number of employees and 55 percent of manufactured exports. See Tain-Jy Chen and others, *Taiwan's Small- and Medium-Sized Firms' Direct Investment in Southeast Asia* (Taipei: Chung-Hua Institution for Economic Research, April 1995). For an insightful discussion of the fundamental difference firm size makes in the development patterns of Taiwan and South Korea, see Tibor Scitovsky, "Economic Development in Taiwan and South Korea, 1965–81," in Lawrence Lau, ed., *Models of Development: A Comparative Study of Economic Growth in South Korea and Taiwan* (San Francisco: Institute for Contemporary Studies, 1986), pp. 135–208.

19. Barry Naughton, *Growing Out of the Plan: Chinese Economic Reform, 1978–1993* (Cambridge University Press, 1995), pp. 137–69.

20. For an engrossing account of the dense network of contracting and subcontracting relationships that allow the Taiwan shoe industry to respond rapidly to changing demand, costs, and fashions, see You-tien Hsing, *The Taiwan Connection: Making Capitalism in China* (Oxford University Press, 1997).

21. Donald B. Keesing and Sanjaya Lall, "Marketing Manufactured Exports from Developing Countries: Learning Sequences and Public Support," in Gerald K. Helleiner, ed., *Trade Policy, Industrialization, and Development: New Perspectives* (Oxford: Clarendon Press, 1992), pp. 176–93; Paul Krugman, "Technology and Changing Comparative Advantage in the Pacific Region," in Hadi Soesastro and Mari Pangestu, eds., *Technological Challenge in the Asia-Pacific Economy* (Sydney: Allen and Unwin, 1990), pp. 25–37; and Michael Plummer and Manuel Montes, "Direct Foreign Investment in China: An Introduction," in La Croix, Plummer, and Lee, eds., *Emerging Patterns,* pp. 3–20. The ability of China Circle producers to respond to product specifications in developed-country markets can be traced back to the porcelain trade of the eighteenth century. For an interesting discussion of the development of these capabilities in the 1950s in the context of the export of rattan and plastic flowers from Hong Kong, see Matthew Turner, "Hong Kong Design and the Roots of Sino-American Trade Disputes," *Annals of the American Academy,* no. 547 (September 1996), pp. 37–53. Developed-country buyers were also important in the development of the Korean electronics industry, but large Korean firms assumed more export functions more quickly than those in Taiwan or Hong Kong. See Mike Hobday, "East Asian Latecomer Firms: Learning the Technology of Electronics," *World Development,* vol. 23, no. 7 (1995), pp. 1171–93.

22. Dieter Ernst, "Mobilizing the Region's Capabilities? The East Asian Production Networks of Japanese Electronics Firms," in Eileen Doherty, ed., *Japanese Investment in Asia: International Production Strategies in a Rapidly Changing World* (San Francisco: Asia Foundation, 1996), p. 32. See also Robert Z. Lawrence, "How Open Is Japan?" in Paul Krugman, ed., *Trade with Japan* (University of Chicago Press, 1991); and Shujiro Urata, "Changing Patterns of Direct Investment and the Implications for Trade and Development," in C. Fred Bergsten and Marcus Noland, eds., *Pacific Dynamism and the International Economic System* (Washington: Institute for International Economics, 1993), pp. 291–93.

23. Textiles and garments were the first sectors in which large-scale trade began

across the Taiwan Strait. In the mid-1980s the mainland began to export primarily natural fibers and imported synthetic fibers and cloth. "Growing Direct Trade Links Peking with Taipei, Seoul," *Business China*, October 12, 1987, pp. 145–47.

24. Irene Wing-chiu Ng, "Flexible Production and the Creation of Competitive Advantage in an Asian Newly Industrializing Economy: Organization and Dynamics in the Hong Kong Electronics Industry," Ph.D. dissertation, University of California, Los Angeles, 1992; available from University Microfilms International, pp. 234, 237, 302.

25. Kenneth Kraemer and others, "Entrepreneurship, Flexibility, and Policy Coordination: Taiwan's Computer Industry," *Information Society,* vol.12 (Spring 1996), pp. 215–49.

26. Soohyun Chon, "Development of the Semiconductor Industry in Korea: A Stepping Stone to Join the Ranks of Core Countries?" *Competition and Change,* vol. 1, no. 3 (1996), pp. 259–83.

27. *Zhongguo gongye jingji tongji ziliao, 1986* (China industrial economics statistical material) (Beijing: Zhongguo Tongji), p. 49; and *Zhongguo gongye jingji tongji nianjian, 1995* (China industrial economics statistical yearbook) (Beijing: Zhongguo Tongji) p. 214.

28. Shi Liu, "The Development of China's Electronics Industry Has Been Greatly Aided by Foreign Investment" (in Chinese), *Zhongguo Jingji Xinwen* (China economic news), August 19, 1996, p. 18. The percent share given here is relative to all exports; note that the output share shown in table 1-2 is the share of output of large (independent accounting) factories only. Thus, as one would expect, the foreign share of exports is greater than that of output.

29. Moreover, the submicron level was likely to be achieved in only a handful of factories: most of the "champions" had already dropped out of the competition. For Shanghai in 1985, see Denis Fred Simon and Detlef Rehn, *Technological Innovation in China: The Case of the Shanghai Semiconductor Industry* (Ballinger, 1988), p. 121; for 1995 technology and the assessment of delays in technology assimilation, see Ying Huachun and others, "A Golden Opportunity for a Renaissance: Developing Our IC Industry" (in Chinese), *Zhongguo touzi yu jianshe* (China investment and construction), no. 6 (1995), pp. 22–25. Additional information on government policy in the 1980s can be found in Jonathan Pollack, "The Chinese Electronics Industry in Transition," Rand N-2306 (Santa Monica, Calif.: Rand Corporation, May 1985). A good brief discussion is Lo Dic, "The Chinese Electronics Industry: State Industrial Policy and Development" (Hong Kong: CERD Consultants, 1994). Many of China's top leaders emerged from the group that designed electronics industry policy in the 1980s: Li Peng, the current premier, was head of the Electronics Leading Group and a key figure in the policy; Jiang Zemin, the current head of the Communist party, was then Minister of Electronics. See also Detlef Rehn, "Organizational Reforms and Technology Change in the Electronics Industry: The Case of Shanghai," in Denis Fred Simon and Merle Goldman, eds., *Science and Technology in Post-Mao China* (Harvard University, Council on East Asian Studies, 1989), pp. 137–55.

30. The shift in orientation at the Ministry of Electronics is often associated with Hu Qili, a noted reformer, who became minister in the early 1990s. See Saiman Hui and Hilary B. McKowan, "China Computes," *China Business Review*, vol. 20 (September–October 1993), pp. 14–20. On current policy, see Zheng Xinli, ed., *Zhongguo zhizhu chanye zhenxing fanglue* (Strategies for developing China's "pillar industries")

(Beijing: Zhongguo Jihua, 1994), pp. 83–108; and Denis Fred Simon, "From Cold to Hot: China Struggles to Protect—and Develop—a World-Class Electronics Industry," *China Business Review,* November–December 1996, pp. 8–16.

31. *China Customs Statistics, 1996*, vol. 11, p. 8; *1994*, vol. 12, p. 8. The categories refer to SITC revision 3 categories 74 and 75.

32. Naughton, "China's Emergence and Prospects," pp. 309–16.

33. For an illuminating account of the growth of entrepreneurship and technical ability of one firm that emerged from this milieu, see Scott Kennedy, "The Stone Group: State Client or Market Pathbreaker?" George Washington University, Department of Political Science, February 1997.

34. Yang Dejiao, "The Lianxiang [Legend] Group: The Reality of China's Dream of Having a World-Famous Brand" (in Chinese), *Guoji Maoyi* (International trade), no. 5 (1996), pp. 13–15.

35. "Great Wall Group to Lead China's Computer Industry," *China Economic News,* December 30, 1996, p. 10.

36. These relatively successful Chinese computer firms replicate familiar patterns not only in the choice of products and processes to launch initial world market operations but also in corporate form. Both these firms are "network firms," with individual nodes having a high degree of autonomy and various divisions held together by cross-holding links that are often complicated and obscure. In this they resemble characteristics of overseas-Chinese firms, which also apply to many firms based in Hong Kong and Taiwan. See Gary Hamilton and Nicole Biggart, "Market, Culture, and Authority: A Comparative Analysis of Management and Organization in the Far East," *American Journal of Sociology,* vol. 94, Supplement, July 15, 1988, pp. S52–S94; Gary Hamilton, ed., *Business Networks and Economic Development in East and Southeast Asia* (Hong Kong University Center for Asian Studies, 1991); and S. Gordon Redding, *The Spirit of Chinese Capitalism* (New York: Walter de Gruyter, 1990).

37. Terho Uimonen, "Comdex China: Government Hails IT Growth," *IDG Market News Update*, February 27, 1997 [via World Wide Web]; Kelly Her, "Information Industry Racing Ahead," *Free China Journal,* November 29, 1996, p. 3; and Terho Uimonen, "IBM, Partners Build Hard Drive Component Plant in China," *IDG Market News Update*, March 5, 1997. IBM will have an 80 percent stake, while its regular partner in China, Great Wall Computer, will have a 10 percent stake, and Shenzhen Kaifa Technology will hold the remaining 10 percent.

38. Jon Skillings, "China PC Market Hits 2M Units with No Clear Leader," *IDG China Market New Update*, February 12, 1997; and Skillings, "In Asia, Japan Buys More PCs, but China Buys More Quickly," *IDG China Market New Update*, February 18, 1997 [via World Wide Web].

39. Shi Liu, "China's Penetration Rate of Telephones up to 5.8 Percent," *China Economic News,* December 9, 1996, p. 5.

40. "China's 1995 Top 500 FFEs Ranking in Sales," *China Economic News*, Supplement, December 12, 1996, p. 1. Motorola's sales rank it as the number two foreign-invested firm in China, after Shanghai Volkswagen.

PART I

CHAPTER TWO

Hong Kong and the Economic Integration of the China Circle

Yun-wing Sung

THE CHINA CIRCLE comprises Hong Kong, Macao, Taiwan, and mainland China. There are three concentric layers of the China Circle, with the Hong Kong–Guangdong economic nexus as the core; Hong Kong, Guangdong, Fujian, and Taiwan, as the inner layer, and Hong Kong, Taiwan, and China as the outer layer. The pivot for the integration of the China Circle is Hong Kong, and in this chapter I examine its role in each of the three layers of the Circle.[1] Since the inauguration of China's open policy and economic reforms in 1979, intense trade and investment flows have developed in the China Circle, making it a dynamic region with a substantial impact on world trade and investment.

Table 2-1 shows the basic economic indicators of the China Circle. The GDP of Taiwan is 1.8 times that of Hong Kong. China's 1995 exports of $148.8 billion already surpassed Taiwan's exports of $85 billion and also vastly surpassed Hong Kong's domestic exports (that is, exports made in Hong Kong) of $30 billion.* Hong Kong's total

Research for this chapter was supported by a research grant from the Research Grants Council of Hong Kong.

*All dollar amounts are U.S. dollars unless otherwise indicated.

Table 2-1. *Basic Indicators for Greater China, 1995*

Indicators	*Hong Kong*	*Taiwan*	*Macao*[a]	*China*	*Guangdong*	*Fujian*
Area (square kilometers)	1,068	35,961	19	960,000	177,901	121,400
Population (millions)	6.2	21.3	0.4	1,211.2	66.7	32.3
GDP (US$billions)	142.0	257.2	6.5	691.4	65.1	26.3
Per capita GDP (US$)	23,019	12,490	15,878	571	949	814
GDP growth rate (percent)	4.6	6.3	4.0	10.2	15	15
Exports (US$billions)	172.3[b] (30.0)[c]	111.7	1.87	148.8	56.6	7.9

Sources: Data for China, Guangdong, and Fujian are obtained from the respective Statistical Yearbooks (Beijing: China Statistical Publishing House, 1996). Data for Hong Kong are obtained from the *Hong Kong Monthly Digest of Statistics* (Hong Kong: Census and Statistics Department, 1996). Data for Taiwan are taken from the *Monthly Bulletin of Statistics of the Republic of China* (Taipei: Directorate-General of Accounting, Budget, and Statistics, 1996). Data for Macao are taken from the *Asia Yearbook, 1996* (Hong Kong: Far Eastern Economic Review, 1996).

a. Data are for 1994.

b. Total exports (including re-export).

c. Domestic exports.

exports (that is, including re-exports) of $172 billion still exceeded China's exports, but only because Hong Kong was re-exporting Chinese products to third countries and third-country products to China. In other words, Hong Kong is China's gateway to the world in commodity trade. Guangdong's 1995 exports of $56.6 billion already surpassed Hong Kong's domestic exports of $30 billion and was on a par with the exports of Thailand.

The trade flows generated by Hong Kong investment in Guangdong were huge. Most of Hong Kong's outward processing activities in China were in Guangdong, and the bulk of their output was imported into Hong Kong for re-export to third countries. In 1995 Hong Kong's re-exports of goods made in Guangdong under outward processing contracts were estimated to be nearly $63.7 billion, which was more than double the domestic exports of Hong Kong and also exceeded Thailand's exports of $56.7 billion. Guangdong accounted for roughly 30 percent of the cumulative foreign direct investment in China. In 1995 FDI in Guangdong was over $10 billion, more than four times

Thailand's total of $2.3 billion. It should be noted that Fujian was a distant second to Guangdong in economic strength. Fujian's 1995 exports and GDP were only 14 percent and 40 percent, respectively, of Guangdong's. A majority of Taiwanese originated from Fujian, and Taiwan accounted for the bulk of the FDI in that province. However, the prime destination of Taiwanese investment in China is still Guangdong, because of its economic dynamism.[2]

The China Circle, or Greater South China, was the first and most successful subregional economic zone in East Asia. As East Asian countries liberalized their economies in the 1980s, many subregional economic zones emerged because of geographic and market forces. Trade and investment flows grew among geographically contiguous but politically separate border areas, taking advantage of the complementarities in factor endowment and technological capacity among countries at different stages of economic development.[3] These subregional economic zones are variously called transnational export processing zones, natural economic territories,[4] and growth triangles (ASEAN terminology), and they include the growth triangle involving Singapore, the Johor state in Malaysia, and Batam Island in Indonesia.[5]

The impetus of the integration of the China Circle came primarily from the economic liberalization of China and secondarily from the economic liberalization of Taiwan. Despite economic liberalization there are still many barriers to the economic integration of China, Hong Kong, and Taiwan: foremost are the remnants of central economic planning in China and Taiwan's ban on direct business links with the mainland. Furthermore, no institutional framework is coordinating the economic integration of the trio. However, geographic and cultural proximity and the huge gains from economic complementarity have overcome the many barriers to economic interactions. Private initiative and market forces have led to intense trade and investment flows among the trio despite the lack of an institutional framework.

From 1991 to 1992 foreign direct investment in China jumped from $4.4 billion to $11 billion, and China became by far the largest recipient of FDI among developing countries, with a share of 22 percent of the total. FDI in China continued to rise rapidly, reaching $27.5 billion in 1993, $33.8 billion in 1994, and $37.5 billion in 1995, accounting for 38 to 39 percent of total FDI in developing countries. From 1993 to 1995 FDI in China exceeded by far the FDI in the entire ASEAN group, and China was second only to the United States in inward FDI.

Overview of the Integration of the Trio

Before China's opening, economic ties linking the mainland and Hong Kong were quite strong, but the relationship was asymmetric. Hong Kong was open to China's exports and investment, but China was closed to Hong Kong's exports and investment. In the 1960s Hong Kong was the foremost market for the mainland. China's trade surplus with Hong Kong was about one-fifth of its total exports, and China used the hard currency thus earned to finance imports of grain, industrial raw materials, and capital goods from developed countries. With the inauguration of China's economic reforms and open policy in 1978, the economic relationship between the mainland and Hong Kong became multifaceted and more balanced. Since the late 1980s the economies of the mainland and Hong Kong have become highly integrated, and the mainland and Hong Kong are now each other's foremost trade and investment partners.

Despite the absence of official ties, economic integration proceeded rapidly between the mainland and Taiwan, largely through use of the efficient intermediary services of Hong Kong. In 1991 Taiwan surpassed Hong Kong and the United States to become the second largest supplier of goods (after Japan) to the mainland and also surpassed the United States and Japan to become the second largest investor (after Hong Kong) on the mainland. In 1992 China surpassed Japan to become the second largest market (after the United States) for Taiwan's exports.

Hong Kong and south China are much more tightly integrated than Taiwan and south China as a result of both geography and Taiwan's policy of no direct business links with China. There is of course no land bridge connecting Taiwan with the mainland. Unlike Taiwan, Hong Kong can fully exploit vertical complementarity by using trucks carrying semimanufactures to its subsidiaries across the border. For Taiwan, investing in south China is not that different from investing in Southeast Asia in terms of labor costs, transportation costs, and turnaround time, although south China has the advantage of cultural proximity.

Integration via Cultural Affinity despite Lack of Institutions

The three important institutional barriers to economic integration most often cited are tariffs, controls on factor movements, and exchange risks. On all three counts, barriers to economic integration among the trio are very high. For instance, even though China will resume sovereignty over Hong Kong in 1997, it is specified in the Sino-British agreement on Hong Kong that Hong Kong will remain a separate customs territory with its own currency. Migration from China to Hong Kong will be strictly controlled. Even after 1997 Hong Kong and the mainland will be less institutionally integrated than Greece and Ireland, which are both members of the European Union (EU) and allow complete freedom of movement of goods and factors between countries. Members of the European Monetary System within the EU are even more closely integrated because of their pegged exchange rates. Since China is not a member of the World Trade Organization (WTO) and the Chinese currency is not fully convertible, Hong Kong is institutionally more closely integrated with most other economies than with China.

Although economic theory concentrates on tariffs, controls on migration, and exchange integration, the effect of geographic and cultural distances may be even more important. Hong Kong is only a thirty-minute train ride from China, and people in Hong Kong had their ancestral roots in Guangdong, the primary site of Hong Kong's investment in China. As mentioned, people in Taiwan have their roots in Fujian and account for the bulk of investment there. These geographic and cultural proximities enable businesses to evade formal barriers to trade and investment. Tariffs can be evaded through smuggling, and smuggling is rampant from Hong Kong and Taiwan to China. The movement of people from Hong Kong and Taiwan to China is relatively free even though movement in the other direction is highly controlled. Even so, illegal immigrants from the mainland are common in Hong Kong and Taiwan, for the labor markets in the two economies are extremely tight. Although the Chinese yuan is not convertible, Hong Kong currency circulates widely (and unofficially) in Guangdong, especially in the Shenzhen special economic zone (SEZ). The Hong Kong government estimates that 22 percent to 25 percent of the total supply of the Hong Kong currency (roughly HK$17 billion) circulates in China.[6] A gray market for yuan also existed in Hong Kong for some time. The gray market became an open market in 1993, when China officially permitted visitors to bring 6,000 yuan outside or into China. Many Hong Kong tourists' shops now accept payment in yuan.

Unilateral Policy Changes

Unilateral policy changes are also important in the integration of the trio. China tailored its initial open-door policies to forge closer links with Hong Kong and Taiwan. Guangdong and Fujian were given special policies in 1979 that vastly increased their autonomy, including managing foreign trade and investment. The first four SEZs were all in Guangdong and Fujian and targeted Chinese populations in Hong Kong, Taiwan, and Southeast Asia. Taiwanese businesses enjoy special concessions in China over other overseas business. China started to woo Taiwan in 1980 by granting Taiwanese goods lower taxes and less stringent import controls. A 1988 State Council decree also gave Taiwanese investors favorable treatment over other foreign investors.[7] Local authorities tend to give Taiwanese investors more favorable treatment by faster approval or better support services.

Although the mainland is more open to Taiwan than to any other economy, Taiwan is less open to the mainland. But Taiwan's import controls on Chinese products have been gradually liberalized since 1987. The number of items allowed to be indirectly imported increased from 29 items in 1987 to 90 items in 1989, 155 items in 1990, and 1,654 items by the end of 1993.[8] On July 1, 1996, Taiwan liberalized further by changing from positive licensing to negative licensing—that is, imports will be freely allowed unless they fall within the controlled list—though positive licensing has been retained for agricultural products.

In July 1987 Taiwan eased its foreign exchange controls, and Taiwanese businesses started to invest indirectly on the mainland through subsidiaries established in Hong Kong or elsewhere. In November 1987 Taiwan allowed its citizens to visit their mainland relatives, and Taiwanese visitors to the mainland soared. In October 1989 Taiwan promulgated regulations sanctioning indirect trade, investment, and technical cooperation with China. Taiwan's policy requires that all trade, investment, and visits be conducted indirectly (that is, via Hong Kong or other third places). Taiwan still prohibits investment from China, although it is reported that mainland investment in Taiwan exists through overseas subsidiaries.

Although Hong Kong businesses receive no favorable concessions in China over other overseas businesses, Hong Kong businesses have significant advantages because of geographic proximity and kinship links. Hong Kong investors obtain favorable concessions from local authorities in Guangdong as a result of the kinship network, and it is easier for Hong Kong Chinese to visit the mainland, since visas are not required. China is thus more open to Hong Kong than to other economies, and

Hong Kong is open to the whole world, including China. But it should be noted that Hong Kong's controls on visitors from the mainland are stricter than controls on other visitors because of the fear of illegal immigrants from the mainland. In cooperation with Hong Kong, Beijing also imposes strict controls on visits to Hong Kong.

China is planning to abolish special favors for Taiwanese and overseas-Chinese investors as part of its reform package to gain entry into the WTO. However, Hong Kong residents and Taiwanese probably will continue to enjoy simpler border formalities and special informal treatment from local authorities in Guangdong and Fujian.

Trade and Investment

Table 2-2 shows China's contracted inward investment by source. Hong Kong is by far the largest investor in China, and Taiwan is a distant second; the United States and Japan are third and fourth. The large share of Hong Kong in China's investment conceals an important intermediary role for Hong Kong. In China's statistics, investment from Hong Kong includes the investment of the subsidiaries of foreign companies incorporated in Hong Kong. Many multinational companies like to test the Chinese investment environment through investments from their Hong Kong subsidiaries, because Hong Kong has the required expertise and is the foremost center for China's trade and investment. Chinese enterprises also invest in China from their Hong Kong subsidiaries to take advantage of preferences given to foreign investors. There is no reliable estimate on the amount of Chinese capital "round tripping" via Hong Kong.

Hong Kong's Investment in China

Hong Kong's investment in China is diversified, ranging from small-scale labor-intensive operations to large-scale infrastructure projects. From 1979 to 1995 Hong Kong accounted for more than 80 percent of the utilized FDI in Guangdong, and Guangdong accounted for more than 40 percent of Hong Kong's FDI in China. Hong Kong's industrial investment in Guangdong transformed Hong Kong manufacturing and its entire economy. Hong Kong manufacturing firms employ up to 3 million workers in Guangdong, whereas in 1995 the manufacturing

Table 2-2. *Contracted Foreign Investment in China by Selected Countries, 1979–95*
Millions of U.S. dollars; numbers in parentheses are percents of total

Item	*1979–90*	*1991*	*1992*	*1993*	*1994*	*1995*	*1979–95*
National total	45,244	12,422	58,736	111,967	83,088	91,282	402,739
	(100)	(100)	(100)	(100)	(100)	(100)	(100)
Hong Kong	26,480	7,531	40,502	74,264	47,278	42,111	238,166
	(58.5)	(60.6)	(69.0)	(66.3)	(56.9)	(46.1)	(59.1)
Taiwan	2,000	1,392	5,548	9,970	5,397	5,849	30,184
	(4.4)	(11.2)	(9.4)	(8.9)	(6.5)	(6.4)	(7.5)
United States	4,476	555	3,142	6,879	6,027	7,471	28,540
	(9.9)	(4.5)	(5.3)	(6.1)	(7.3)	(8.2)	(7.1)
Japan	3,662	886	2,200	3,015	4,457	7,592	21,812
	(8.1)	(7.1)	(3.7)	(2.7)	(5.4)	(8.3)	(5.4)

Sources: Data before 1995 come from *Almanac of China's Foreign Relations and Trade* (Beijing: China Economics Publishing House), various issues. Data for 1995 pertain to foreign direct investment (over 99 percent of foreign investment) and are taken from *China Economic News* (Beijing: *Economic Daily,* External Services Division, no. 22, June 17, 1996).

labor force in Hong Kong fell from a peak of 905,000 in 1984 to 386,000. By moving the labor-intensive processes to Guangdong, Hong Kong can concentrate on more skill-intensive processes such as product design, sourcing, production management, quality control, and marketing. Hong Kong manufacturing was able to achieve a very high rate of labor productivity growth. The expansion of exports from processing operations in Guangdong also increased the demand for Hong Kong's service industries, including entrepôt trade, shipping, insurance, business services, and financial services. Both the Hong Kong manufacturing sector and the Hong Kong economy became increasingly service oriented—in short, becoming the economic capital of an industrialized Guangdong.

Before Deng Xiaoping's tour of southern China in early 1992 in support of economic reforms, the largest corporations in Hong Kong were not active investors in China, although small and medium-size Hong Kong enterprises, especially Hong Kong's labor-intensive manufacturing firms, invested in China in droves. Deng's tour stimulated a wave of investment by major Hong Kong companies, including listed companies such as Cheung Kong, Hutchison-Whampoa, Sun Hung Kai Properties, New World, and Kowloon Wharf, in projects ranging from real estate to infrastructure and commerce. At about the same time small investors also purchased large numbers of pre-sale flats in Guangdong as speculative investment. As a result of these developments, Hong Kong's already high share in China's contracted foreign investment reached a peak of 69 percent in 1992. This showed that Hong Kong investors were sensitive to investment opportunities in China; they were one step ahead of other investors. As other investors jumped on the bandwagon of China's boom, Hong Kong's extraordinarily high share declined to 46 percent in 1995 (table 2-2).

China's Investment in Hong Kong

In the 1990s China has become an important investor in the world. According to the International Monetary Fund and United Nations Conference on Trade and Development,[9] from 1990 to 1994 China's average annual outward FDI reached $2.4 billion, more than that of Latin America and the Caribbean combined. China's investment in Hong Kong is very diversified, covering nearly all sectors of the Hong Kong economy, namely, banking, insurance, entrepôt trade, shipping, aviation, real estate, and manufacturing. China's investment strengthens the ties of Hong Kong to China and enhances the position of Hong Kong as the gateway to China.

Unfortunately, the precise amounts are difficult to estimate. In May 1996 the Hong Kong government published the first surveys (1993 and 1994) of external investment in Hong Kong's nonmanufacturing sectors, and these give the first precise data on external investment in Hong Kong.[10] At the end of 1994 China was the third external investor in Hong Kong, with a share of net assets of 18.4 percent, after the United Kingdom (28.2 percent) and Japan (20.7 percent). The United States was a distant fourth, with a share of 12.0 percent. Net assets of Chinese investment totaled $17.2 billion at the end of 1994.

Before the release of the survey results, it was widely reported in the press that China has replaced the United Kingdom and Japan as the foremost external investor in Hong Kong. Such reports have some truth, since many local governments in China invest in Hong Kong without Beijing's approval, and these companies are usually incorporated under the names of their Hong Kong relatives. China's investment in Hong Kong is thus probably understated in the survey of the Hong Kong government. According to my estimation, cumulative FDI of China in Hong Kong should have reached $20 billion by the end of 1994, just surpassing that of Japan.[11] Moreover, the United Kingdom has a long history of investment in Hong Kong; its new investments mainly come from retained earnings, while Chinese investment in Hong Kong probably involves more net inflows of capital.

Hong Kong's investment in China appeared to be substantially larger than the reverse flow. Hong Kong's utilized direct investment in China amounted to $60 billion at the end of 1994, exceeding considerably China's estimated investment in Hong Kong of $20 billion. Hong Kong's investment in China is significantly exaggerated because it includes the investment from the subsidiaries of other multinationals incorporated in Hong Kong. Moreover, officials in planned economies tend to magnify economic performance (the "success indicators" problem). From anecdotal evidence, it is known that Hong Kong investors often overstate the value of their investments in China, with the connivance of local officials. For example, Hong Kong manufacturers tend to put a high value on the outdated machinery they move to China.[12]

As China continues to liberalize its foreign exchange controls, more and more Chinese capital will most likely flow to Hong Kong through official as well as unofficial channels. It is natural for Chinese enterprises and investors to move their capital to Hong Kong, for Hong Kong has stricter protection of property rights than China does, and the funds can also be used much more flexibly in Hong Kong. In the long run Chinese investment in Hong Kong may rival Hong Kong's investment in China.

China's Trade with Hong Kong

Hong Kong and China are often considered each other's foremost trading partner. While this statement is technically true, it is misleading, because it includes China's trade with third countries via Hong Kong (Hong Kong's entrepôt trade). Here I use mainly Hong Kong statistics in discussing Hong Kong–China trade, because China's statistics fail to distinguish between China's trade with third countries via Hong Kong and China's trade with Hong Kong itself.

In trade statistics, exports are classified by country of destination, whereas imports are classified by country of origin. In U.S.-China trade, both countries regard their exports to each other through Hong Kong as exports to Hong Kong, understating exports to each other. Imports are not understated, since they are traced to the country of origin. Both countries thus overstate their bilateral trade deficits or understate their bilateral surpluses. For example, in 1992 U.S. statistics claimed a deficit of $18 billion in its trade with China, whereas China claimed a trade deficit of $306 million with the United States. American statistics are less misleading than those of China, however, because in the early 1990s about two-thirds of China's exports to the United States were re-exported through Hong Kong, whereas the corresponding figure for U.S. exports to China was only about 20 percent.[13]

Hong Kong's imports of Chinese goods in 1995 totaled $67.3 billion, 92 percent re-exported to third countries and only 8 percent retained in Hong Kong. Although China was by far the foremost supplier of Hong Kong's re-exports, it supplied only 9.6 percent of Hong Kong's retained imports in 1994, ranking third after Japan and the United States. China has been unable to capture the higher end of Hong Kong's market, dominated by Japan. Given the increasing affluence of Hong Kong and the Japanese dominance in vehicles, capital goods, and quality consumer durables and consumer goods, the future of Chinese products in Hong Kong is not bright.

Hong Kong was the largest final market (that is, excluding Chinese exports through Hong Kong) for Chinese exports in the late 1960s and early 1970s, but the Hong Kong market was overtaken by the Japanese market and the U.S. market in 1973 and 1987, respectively. In 1995, though the Hong Kong market accounted for 42.7 percent of China's exports, the bulk (39.2 percent) was re-exported, with only 3.5 percent retained in Hong Kong.

The Hong Kong–Guangdong Production Network

Hong Kong's investment in outward processing operations in China, especially in Guangdong, has generated huge trade flows (table 2-3). In 1995 Hong Kong's imports from China involving outward processing amounted to $51.7 billion, or nearly 74 percent of Hong Kong's total imports from China. Guangdong clearly accounted for the bulk of the outward processing operations in China, since Hong Kong's imports from Guangdong related to outward processing accounted for 93 to 95 percent of the imports from China involving outward processing from 1989 to 1995.

Most of Hong Kong's imports involving outward processing were further processed or packaged in Hong Kong for exports to third countries. If the processing substantially changes the form or nature of the

Table 2-3. *Hong Kong Trade Involving Outward Processing Operations in China, 1989–95*

Amounts in millions of U.S. dollars

Item	*Exports to China*			*Imports from*		
	Domestic exports	*Re-exports*	*Total*	*China*	*Guangdong*	*Re-exports of China origin*
1989						
Amount	4,098	5,757	9,855	14,562	13,601	n.a.
Percent of total[a]	(76.0)	(43.6)	(53.0)	(58.1)	n.a.	n.a.
1990						
Amount	4,676	7,125	11,800	18,629	17,592	n.a.
Percent of total[a]	(79.0)	(50.3)	(58.8)	(61.8)	n.a.	n.a.
1991						
Amount	5,195	9,466	14,661	25,400	24,011	28,497
Percent of total[a]	(76.5)	(48.2)	(55.5)	(67.6)	n.a.	(74.1)
1992						
Amount	5,719	12,578	18,297	32,566	30,335	38,733
Percent of total[a]	(74.3)	(46.2)	(52.4)	(72.1)	n.a.	(78.3)
1993						
Amount	5,835	14,870	20,706	38,160	35,617	47,122
Percent of total[a]	(74.0)	(42.1)	(47.9)	(73.8)	n.a.	(80.8)
1994						
Amount	5,429	18,015	23,444	45,925	43,372	54,677
Percent of total[a]	(71.4)	(43.3)	(47.7)	(75.9)	n.a.	(82.0)
1995						
Amount	5,673	22,456	28,130	51,650	49,068	63,658
Percent of total[a]	(71.4)	(45.4)	(49.0)	(74.4)	n.a.	(82.2)

Sources: *Hong Kong External Trade* (Census and Statistics Department, Hong Kong, various issues).
n.a. Not available.
a. Proportion of outward processing trade in total.

product, then it is classified as Hong Kong's domestic exports, that is, exports of goods made in Hong Kong. Otherwise it is classified as Hong Kong's re-exports. Since 1991 Hong Kong's re-exports of China origin involving outward processing exceeded Hong Kong's domestic exports. By 1995 Hong Kong's re-exports of China origin involving outward processing were $63.7 billion, or more than double that of Hong Kong's domestic exports.

Adjusting for the Biases of Outward-Processing Trade. In view of the importance of outward processing trade, one needs to interpret Hong Kong's trade statistics with care, giving recognition to the special characteristics of outward processing trade. For instance, the share of China in Hong Kong's trade is biased upward, because Hong Kong's domestic exports of semimanufactures to China are re-imported into Hong Kong after processing in China and may even be re-exported to China and re-imported into Hong Kong a few more times before final export to third countries. Moreover, the overall growth rate of Hong Kong's exports also gives a misleading impression. From 1978 to 1995 total Hong Kong exports grew at the average rate of 17 percent a year, an extremely high rate of growth.

Table 2-4 tries to correct for the biases introduced by outward processing trade in Hong Kong's export value, growth rates, and the market composition of exports.[14] The first row shows the value, growth rate, and market composition of Hong Kong's domestic exports. Domestic exports to China grew rapidly because Hong Kong firms supply their subsidiaries in China with materials and components made in Hong Kong. Hong Kong's domestic exports to China grew from negligible amounts to $7.9 billion in 1995, and China surpassed the United States as Hong Kong's foremost market, attracting 27 percent of domestic exports.

The second row in the table shows corresponding statistics for Hong Kong's domestic exports of final goods (that is, total domestic exports less domestic export of semimanufactures to China for export processing). The value and growth rate of domestic exports of final goods are naturally less than those of domestic exports. More important, for domestic exports of final goods, the decline in the market shares of the United States and the EU are less dramatic, and the United States is still Hong Kong's foremost market.

The third row in the table shows the exports of final goods of Hong Kong firms in both Guangdong and Hong Kong. That is equal to the total of domestic exports of final goods and Hong Kong's re-exports of goods from outward processing operations in Guangdong, which in turn is 95 percent of Hong Kong's re-exports of goods from outward processing operations in China.[15] The value and growth rate of exports of final goods of Hong Kong firms are of course much higher than those of domestic exports. More important, the United States is once again the foremost market. The share of the U.S. market declined only

Table 2-4. *Hong Kong Exports and Exports of Hong Kong Firms in Hong Kong and Guangdong, 1978, 1995*

	Value (US$millions)		*Growth rate (percent)*	*Market shares (percent)*											
				United States		*China*		*Japan*		*European Union*		*Germany*		*United Kingdom*	
Item	*1978*	*1995*	*1978–95*	*1978*	*1995*	*1978*	*1995*	*1978*	*1995*	*1978*	*1995*	*1978*	*1995*	*1978*	*1995*
1. Domestic exports	8,690	28,945	7.5	37.2	26.4	0.2	27.4	4.6	5.1	26.7	17.3	10.9	5.3	9.5	4.7
2. Domestic exports of final goods[a]	8,690	24,272	6.2	37.2	32.6	0.2	10.5	4.6	6.3	26.7	21.3	10.9	6.5	9.5	5.8
3. Exports of final goods of Hong Kong firms in Hong Kong and Guangdong[b]	8,690	84,747	14.3	37.2	33.5	0.2	3.0	4.6	8.5	26.7	22.9	10.9	6.7	9.5	5.0
4. Total exports	11,507	173,750	17.3	30.3	21.7	0.5	33.3	7.7	6.1	21.7	14.9	8.6	4.3	7.4	3.2
5. Total exports of final goods[c]	11,507	145,620	16.1	30.3	25.9	0.5	20.4	7.7	7.3	21.7	17.8	8.6	5.1	7.4	3.8

Sources: *Hong Kong External Trade*, various issues.

a. Domestic exports of final goods = domestic exports – domestic exports to China involving outward processing.

b. Exports of final goods of Hong Kong firms in Hong Kong and Guangdong = domestic exports of final goods + 0.95 × re-exports of China origin (except to China) involving outward processing; 0.95 is the share of Guangdong in Hong Kong's imports from China related to outward processing. The market composition of Hong Kong's re-exports of Guangdong origin involving outward processing is not available, and it is assumed to be the same as that of HK's re-exports of China origin. For 1995 the former was 78.1 percent of the latter.

c. Total exports of final goods = total exports – total exports to China involving outward processing.

slightly, from 37.0 percent in 1979 to 33.5 percent in 1995. The share of the Chinese market was only 3 percent, though that is an understatement, because Hong Kong manufacturers in Guangdong can sell part of their output directly in China's domestic market.

The fourth row in table 2-4 shows Hong Kong's total exports, which is the total of domestic exports and re-exports. In 1995 Hong Kong's total exports to China were $57.9 billion, 86 percent of which were re-exports of third-country goods to China and 14 percent Hong Kong's domestic exports. The trend of total exports is dominated by that of re-exports. Both the value and growth rate of total exports are very high, and China was the foremost market of Hong Kong's total exports in 1995. The share of outward processing trade in total exports to China declined from 59 percent in 1990 to 49 percent in 1995 (table 2-3). The decline again occurred because of China's import liberalization and the resulting rise in imports of goods not related to outward processing. Unlike the share of domestic exports, the share of Hong Kong's re-exports in China's imports rose continuously, from 2 percent in 1979 to 40 percent in 1995. This shows that, with China's import liberalization, increasingly large varieties and amounts of goods of third countries were re-exported via Hong Kong to China. This result can be attributed to Hong Kong's efficiency as an entrepôt. In 1995 China imported 46.5 percent of its total imports from Hong Kong (including Hong Kong's domestic exports and re-exports).

Finally, the fifth row in the table shows Hong Kong's total exports of final goods, which is Hong Kong's total exports less Hong Kong's total exports to China related to outward processing. However, re-exports not related to outward processing are included. The adjustment is intended to avoid double counting. Even though China has replaced the United States as Hong Kong's foremost market for total exports, the United States is still the top market for total exports of final goods. Even so, the 20 percent market share of China in total exports of final goods in 1995 was still quite high.

To summarize, if one nets out trade in semimanufactures between Hong Kong and China, the United States and the EU are still the largest market for both Hong Kong products and the exports of Hong Kong firms in Hong Kong and in Guangdong. With import liberalization in China, however, the PRC is also becoming an important market for final goods.

Table 2-5 shows exports of Hong Kong firms in Hong Kong and Guangdong by commodity. Many labor-intensive industries have a

Table 2-5. *Exports of Hong Kong Firms in Hong Kong and Guangdong, by Commodity, 1995*[a]

Millions of U.S. dollars; numbers in parentheses are percents of total

		Exports of Hong Kong firms produced in		
SITC	*Commodity*	*Hong Kong*[b]	*Guangdong*[c]	*Total*
83	Travel goods and handbags	105 (3.5)	2,869[d] (96.5)	2,974 (100.0)
894	Toys	341 (3.7)	8,800 (96.3)	9,141 (100.0)
76	Telecommunications and sound recording equipment	1,369 (13.0)	9,172 (87.0)	10,541 (100.0)
899	Miscellaneous manufactures	305 (15.3)	1,695[d] (84.7)	2,000 (100.0)
69	Metal manufactures	606 (30.7)	1,365[d] (69.3)	1,971 (100.0)
65	Textiles	1,814 (33.1)	3,660[d] (66.9)	5,474 (100.0)
75	Office machines and automatic data processing machines	2,309 (43.6)	2,981[d] (56.4)	5,290 (100.0)
77	Electrical machinery and appliances	4,122 (47.2)	4,611[d] (52.8)	8,733 (100.0)
84	Clothing	9,540 (52.8)	8,524 (47.2)	18,064 (100.0)
885	Watches and clocks	1,761 (57.7)	1,289[d] (42.3)	3,050 (100.0)
	Subtotal	20,511 (31.3)	44,966 (68.4)	65,477 (100.0)
	All commodities	29,945 (33.1)	60,475 (66.9)	90,420 (100.0)

Source: Census and Statistics Department of Hong Kong.

a. Commodities are ranked in descending order of the shares of exports produced in Guangdong.

b. Hong Kong's domestic exports.

c. Re-exports of Guangdong origin involving outward processing (taken to be 94.4 percent of re-exports of China origin involving outward processing).

d. Data on re-exports of China origin involving outward processing are not available for these commodities. They are assumed to be equal to $0.95 \times 0.718 \times$ Hong Kong's re-exports of China origin of the respective commodities; 0.718 is the proportion of outward processing trade in China's re-exports of China origin, while 0.95 is the proportion of Guangdong in Hong Kong's outward processing trade with China.

major share of their exports produced from outward processing operations in Guangdong. These include travel goods and handbags (96.5 percent), toys (96.3 percent), telecommunications and sound recording equipment (87 percent), and miscellaneous manufactures (84.7 percent). The more skill-intensive industries have a smaller proportion of their exports produced in Guangdong. These include watches and clocks (42.3 percent), electrical machinery and appliances (52.8 percent), and office machines and data processing machines (56.4 percent). Textiles and clothing also have a relatively small proportion of their exports produced in Guangdong, because exports of textiles and clothing are restricted by quota, and Hong Kong has the largest clothing quota in the world.

The Production Network in the Electronics Industry. The electronics industry is Hong Kong's second largest manufacturing industry after clothing. Because Hong Kong's exports of clothing have been restricted by quota since the Lancashire Pact of 1958, pundits have predicted since the 1960s that electronics would surpass clothing as the number one industry in Hong Kong. That, however, did not occur, simply because Hong Kong's electronics industry has largely moved to Guangdong, while the clothing industry is less footloose because of quota restrictions.

The electronics industry comprises at least three industries listed in table 2-5: telecommunications and sound recording equipment, office machines and automatic data processing machines, and watches and clocks. Hong Kong's toy industry also includes a considerable number of electronic toys, but it is difficult to ascertain the precise figure. Here I focus on the three main industries.

For exports produced in Hong Kong in 1995, the total of the three industries was $5.4 billion, which was much less than that of clothing ($9.5 billion). But for exports produced by Hong Kong firms in Guangdong in 1995, the total of the three industries was $13.4 billion, which vastly exceeded that of clothing ($8.5 billion). The sum of these figures (which equals exports produced by Hong Kong firms in both Hong Kong and Guangdong in 1995, the category defined in table 2-4) equals $18.9 billion for electronics, which slightly exceeded the $18.1 billion for clothing. In other words, for Hong Kong firms in the China Circle, exports of electronics have already surpassed clothing.

Hong Kong's level of technology in electronics, however, is lower than that of the other East Asian newly industrializing economies. Hong Kong excels in producing fashionable consumer electronics characterized by rapid changes in design because it has a flexible production network dominated by small firms. Hong Kong had been the world's leading exporter of radios, calculators, electronic watches, and electronic games. It has been able to produce some sophisticated new products, including electronic products for industrial and commercial use. These include photocopiers, electronic typewriters, video telephones, cellular telephones, facsimile machines, compact disk players, LCD projection panels, and programmable controls. Hong Kong also has some strength in middle-range products in electronic parts and components. But most high-tech parts and components have to be imported.

Since Guangdong's electronics industry is essentially a Hong Kong transplant, its level of technology is not impressive. However, China has a large army of technical personnel and impressive capability in research and development. There is a potential for a tripartite fusion of the mainland's capability in basic technology, Taiwan's capacity in product development, and Hong Kong's ability in product design, financing, and marketing. This fusion will be able to drive the electronics industry of the China Circle forward.

Hong Kong as an Entrepôt and Offshore Trade Broker. Both outward processing and the decentralization of China's foreign trade boosted Hong Kong's trade with China, especially Hong Kong's re-exports of Chinese goods to third countries, and vice-versa. Decentralization vastly increased the number of trading partners and raised the cost of searching for a suitable trade partner. Intermediation emerged to economize these search costs and was channeled to Hong Kong because of its trading efficiency. Conceptually, the re-export margin that Hong Kong earns through entrepôt trade represents export of services. However, such services are embodied in the price of goods sold and thus usually recorded in trade statistics as export of goods rather than export of services.

In commodity trade Hong Kong is also an important center of transshipment for China, as well as a center of so-called offshore trade, in which Hong Kong agents arrange direct shipping between China and third-country destinations. Transshipment means that goods are consigned directly from the exporting country to a buyer in the importing country, though the goods are transported via Hong Kong and the goods usually change vessel in Hong Kong. Chinese goods, for example, are carried by train or coastal vessels to Hong Kong, where they are consolidated into containers for ocean shipping. Unlike re-exports, transshipment is not regarded as part of Hong Kong's trade, since there is no Hong Kong buyer. The government does not collect statistics on the value of transshipments, for they do not go through Hong Kong customs, though their weight and volume are known from cargo statistics. Besides re-exports and transshipment, Hong Kong traders also perform a brokerage role in a substantial portion of China's direct trade. That is called offshore trade because the Hong Kong firm acts as a middleman between offshore production bases and overseas customers, but the goods are not generally transported via Hong Kong.

The best data on transshipment and offshore trade come from surveys conducted on Hong Kong's traders.[16] These show that transshipment and offshore trade are growing in relative importance, being especially significant for goods shipped into China. For goods coming out of China, an overwhelming (but falling) portion of sales of Chinese products was re-exported via Hong Kong. From 1992 to 1995 the share of re-exports declined from 81 percent to 75 percent, while the share of transshipment increased from 11 to 13 percent, and that of offshore trade from 8 to 12 percent. For sales of third-country (non-Chinese) products of the surveyed traders from 1992 to 1995, the share of re-exports dropped markedly, from 52 percent to 46 percent. The share of transshipment declined marginally, from 12.2 percent to 11.6 percent, while the share of direct shipment rose sharply, from 36 percent to 43 percent. For Hong Kong companies selling both Chinese and third-country products, there has clearly been a substitution of direct shipment for re-exports. The reason is that China has liberalized foreign investment in shipping and cargo forwarding, and foreign firms have found it easier to handle Chinese trade through transshipment and direct shipment.

The share of China's trade handled by Hong Kong is appreciably larger than that reflected by re-exports, because the goods handled by Hong Kong traders through transshipment and direct shipment should be included. In 1995 the shares of Chinese exports consumed, re-exported, transshipped, and directly shipped by Hong Kong agents were 3.5 percent, 39.2 percent, 6.8 percent, and 6.0 percent, respectively, totaling 55.8 percent. Hong Kong thus plays an important role in over half of China's exports. On the import side, the shares of China's imports produced, re-exported, transshipped, and directly shipped by Hong Kong agents were 6.6 percent, 39.9 percent, 10.1 percent, and 37.1 percent, respectively, totaling 87.1 percent. Though the figure may be somewhat exaggerated because of biases in estimation, Hong Kong certainly plays an important role in the bulk of China's imports.[17]

Taiwan's Investment in China

Despite the explosive growth of Taiwanese investment on the mainland in recent years, the total stock of contracted Taiwanese investment at the end of 1995 was only 14 percent of that of Hong Kong (table 2-2). This situation shows a considerable potential for further expansion of the Taiwanese share. However, the 1994 "Thousand Islands" incident,[18] as well as China's hostility toward Lee Tenghui in 1995 and 1996, slowed down Taiwan's investment in China.

In the early years Taiwan's investment was largely in small-scale labor-intensive operations producing light manufactures for export. The industries included textiles, shoes, umbrellas, travel accessories, and electronics and were concentrated in Fujian (particularly Xiamen) and Guangdong. But Taiwanese investment has been growing in size and sophistication, with an increasing number of more technology-intensive projects such as chemicals, building materials, automobiles, and electronic products and components. The fields of investment have diversified from manufacturing into real estate, finance, tourism, and agriculture. The location of investment has also spread inland from the coast.

The surge of Taiwanese investment in the mainland raised fears that such investment would lead to the hollowing out of Taiwan's industry and pose a security threat. In July 1990 the Taiwanese government tried to cool down the PRC investment boom by improving the investment environment in Taiwan and steering investment from the mainland to ASEAN countries. Both carrots and sticks were used to prevent Formosa Plastics from implementing a gigantic project to build a naphtha cracking plant in Xiamen. To control the mainland investment boom, Taiwan does not allow investment in industries still competitive in Taiwan. However, these controls are not always observed.

Taiwan's president visited ASEAN countries in early 1994 in an effort to improve the investment environment for Taiwanese investors there. The Taiwanese government is trying to guide the mainland investment boom rather than to reverse it. There are real political differences dividing the mainland and Taiwan, differences that are not going to disappear. But if Taiwan continues to liberalize its relations with the mainland, Taiwanese investment in China will probably rival that of Hong Kong.

China's Trade with Taiwan

As in the case of Hong Kong, Taiwan's investment in China has generated huge trade flows. The bulk of Taiwan's exports to China consists of raw materials, semimanufactures, and machinery supplied to Taiwanese enterprises on the mainland.[19] Semimanufactures were not that important in the mainland's exports to Taiwan because of Taiwan's restrictions on such imports. However, the share of semimanufactures is rising owing to the liberalization of Taiwan's import controls.

The explosive growth of Taiwan's trade with the mainland in the form of Hong Kong re-exports is regularly reported. Hong Kong statistics on re-exports of Taiwanese origin to China, and vice-versa, have often been used by researchers to gauge the magnitude of Taiwan-mainland trade. What is not well known is the existence of substantial direct trade between Taiwan and the mainland. Because of Taiwan's ban on direct trade with the mainland, this trade usually involves switching trade documents. Taiwanese exporters claim that their goods are destined for Hong Kong when the goods leave Taiwan. But on arrival in Hong Kong, the trade documents are switched, and the goods are declared as destined for the mainland. Because the goods are consigned to a buyer in the mainland, they do not go through Hong Kong customs, and no Hong Kong firm can claim legal possession of the goods. Such goods are regarded as transshipment by the Hong Kong government and are not regarded as part of Hong Kong's trade. Such trade is considered "direct trade" here because no third party buys the goods involved for resale. By switching trade documents, the Taiwanese exporters save 0.1 percent of the cost of going through Hong Kong customs. This direct trade also has the advantage of confidentiality, because Hong Kong customs does not keep records.

In direct trade, Taiwan usually records the exports as destined for Hong Kong. Thus Taiwan's "direct exports" to the mainland can be estimated from Taiwan's "missing exports," which are equal to the difference between Taiwan's exports to Hong Kong and Hong Kong's imports from Taiwan after adjusting for the cost of insurance and freight (that is, the difference between c.i.f. [cost, insurance, and freight] and f.o.b. [free on board] prices). This is the trade-partners' statistics technique. Taiwanese firms have estimated that 80 percent of the missing exports go to China.[20] Table 2-6 shows the estimation of the value of Taiwan's missing exports to Hong Kong. From 1975 to 1987, before Taiwan's liberalization of its mainland policy, Hong Kong's imports from Taiwan were 5 percent larger than Taiwan's exports to Hong Kong. This difference represents the cost of insurance

Table 2-6. *Taiwan's Exports to Hong Kong and China, Selected Years, 1988–95*

Millions of U.S. dollars; numbers in parentheses are the percentage distribution of Taiwan's exports to Hong Kong

	Taiwan's exports to Hong Kong						
		Imported into Hong Kong					
				Re-exported		*Missing from Taiwan exports*	
Year	*Total*	*Subtotal*	*Retained for internal use*	*to China*	*elsewhere*		*China's imports*
1988	5,580	5,344	3,209	1,964	171	236	n.a.
	(100.0)	(95.8)	(57.5)	(35.2)	(3.1)	(4.2)	. . .
1989	7,030	6,237	3,376	2,540	321	793	1,856
	(100.0)	(88.7)	(48.0)	(36.1)	(4.6)	(11.3)	. . .
1990	8,570	7,045	3,832	2,875	338	1,525	2,254
	(100.0)	(82.2)	(44.7)	(33.5)	(3.9)	(17.8)	. . .
1991	12,418	9,019	4,354	4,074	591	3,399	3,639
	(100.0)	(72.6)	(35.1)	(32.8)	(4.8)	(27.4)	. . .
1992	15,427	10,722	4,607	5,509	606	4,705	5,881
	(100.0)	(69.5)	(29.9)	(35.7)	(3.9)	(30.5)	. . .
1993	18,455	11,482	4,275	6,596	611	6,973	12,933
	(100.0)	(62.2)	(23.2)	(35.7)	(3.3)	(37.8)	. . .
1995	26,123	15,643	6,269	8,672	702	10,480	14,784
	(100.0)	(59.9)	(24.0)	(33.2)	(2.7)	(40.1)	. . .

Sources: Taiwan's exports to Hong Kong are obtained from the *Monthly Statistics of the Republic of China;* the amount imported into Hong Kong is taken to be Hong Kong's imports from Taiwan (obtained from *Hong Kong Review of Overseas Trade* [Hong Kong: Census and Statistics Department, various years]) less a 5 percent margin to allow for the cost of freight and insurance. Taiwan's exports re-exported via Hong Kong to China and elsewhere are taken to be Hong Kong's re-export of Taiwanese goods to China and elsewhere (obtained from *Hong Kong Review of Overseas Trade*) less a 15 percent margin to allow for the re-export markup and the cost of insurance and freight. Taiwan's re-export retained for internal use in Hong Kong is obtained as a residual. "Direct exports" to China are also obtained as a residual. China's imports from Taiwan are obtained from *China Customs Statistics* (Hong Kong: Economic Information Agency, various years).

n.a. Not available.

and freight, since direct trade was almost nonexistent before the 1987 liberalization. Since 1988 Taiwan's exports to Hong Kong have exceeded Hong Kong's imports from Taiwan (converted to Taiwan's f.o.b prices) by increasingly large margins. Table 2-7 shows Taiwan's direct and indirect trade with China, with Taiwan's direct exports to China taken to be 80 percent of Taiwan's missing exports to Hong Kong. In 1995 Taiwan's total exports to the mainland were $17 billion

Table 2-7. *Taiwan's "Direct" and Indirect Trade with China, 1986–95*[a]
Numbers in parentheses are percent shares of Taiwan's total exports or imports

	Exports (millions of U.S. dollars)			*Imports (millions of U.S. dollars)*			*Transshipment via Hong Kong (tons)*	
Year	*Direct*	*Indirect*[b]	*Total*	*Direct*	*Indirect*[c]	*Total*	*To China*	*From China*
1986	18.4	705	723	0	151	151	1,392	800
	(0.05)	(1.8)	(1.8)	(0.0)	(0.6)	(0.6)	...	...
1987	73.6	956	1,030	0	303	303	1,912	900
	(0.14)	(1.8)	(1.9)	(0.0)	(0.87)	(0.87)	...	...
1988	189	1,964	2,153	0	502	502	8,096	2,595
	(0.31)	(3.2)	(3.5)	(0.0)	(1.0)	(1.0)	...	...
1989	634	2,540	3,174	102	616	718	53,450	6,662
	(0.96)	(3.8)	(4.8)	(0.2)	(1.2)	(1.4)	...	...
1990	1,220	2,875	4,095	352	804	1,156	81,195	12,447
	(1.8)	(4.3)	(6.1)	(0.6)	(1.5)	(2.1)	...	...
1991	2,719	4,074	6,793	655	1,187	1,842	345,700	87,610
	(3.6)	(5.3)	(8.9)	(1.0)	(1.9)	(2.9)	...	...
1992	3,764	5,509	9,273	768	1,184	1,952	872,292	211,026
	(4.6)	(6.8)	(11.4)	(1.1)	(1.6)	(2.7)		...
1993	5,578	6,596	12,174	1,169	1,159	2,328	1,152,363	329,548
	(6.6)	(7.8)	(14.3)	(1.5)	(1.5)	(3.0)	...	...
1994	6,562	7,476	14,028	1,615	1,361	2,976	1,227,000	442,000
	(7.1)	(8.0)	(15.1)	(1.9)	(1.6)	(3.5)	...	...
1995	8,384	8,672	17,056	2,130	1,657	3,787	1,718,000	557,000
	(7.5)	(7.8)	(15.3)	(2.1)	(1.6)	(3.7)	...	...

Sources: Data on indirect trade: *Annual Review of Hong Kong External Trade* (Census and Statistics Department, various issues); data on direct trade: see text; data on transshipment: *Hong Kong Shipping Statistics* (Census and Statistics Department, various issues).

a. See the text for the explanation of "direct" trade.

b. Taiwan's indirect exports are taken to be Hong Kong's re-exports to China of Taiwan origin less a 15 percent margin to allow for the re-export markup and the cost of insurance and freight.

c. Taiwan's indirect imports are taken to be Hong Kong's re-exports to Taiwan of China origin plus a 5 percent margin to allow for the cost of insurance and freight.

(15.3 percent of Taiwan's exports), divided roughly equally between direct and indirect exports.

Unlike exports, Taiwan's imports from the mainland are restricted to selected commodity categories. Prohibited mainland goods are imported into Taiwan with fake country-of-origin certificates, which can be obtained in Thailand, for example, for a mere $100.[21] Thus one cannot estimate Taiwan's direct imports from the mainland as the difference between Taiwan's imports from Hong Kong and Hong Kong's exports to Taiwan. In table 2-7, Taiwan's direct imports from China before 1993 were obtained from China's exports to Taiwan (converted to Taiwan's c.i.f. prices), as recorded in China customs statistics. This method does not work after 1993, though.[22] Taiwan's direct imports since 1993 were estimated from the value of the weight of Hong Kong's transshipment of mainland goods to Taiwan and an estimate of the value per ton of such transshipment. The latter is obtained from 1992 data (dividing Taiwan's 1992 direct imports by the weight of transshipment) and is then adjusted for inflation by using the unit value index of Hong Kong's imports from China. This gives an estimate of direct imports of $2.1 billion in 1995, which exceeded the indirect imports of $1.7 billion by a large margin. Between 1991 and 1995 Hong Kong re-exports of mainland goods to Taiwan stagnated while transshipment of mainland goods to Taiwan continued to soar (table 2-7). Evidently a substitution of direct for indirect trade is taking place.

Taiwan has a massive surplus in its commodity trade with the mainland, partly because of Taiwan's policy of only importing selective commodity items from the mainland and partly because of China's lack of competitiveness in producing items demanded in Taiwan. However, Taiwan has large deficits with the mainland in tourism, gifts and remittances, and investment. The payments balance across the Taiwan Strait is thus more even. Moreover, intra-industry trade is expected to develop rapidly with the surge of Taiwanese investment on the mainland and the further liberalization of Taiwan's controls on imports from the mainland.

Taiwan-PRC trade has grown extremely fast and is now substantial. In 1992 the mainland surpassed Japan to become the second largest market for Taiwan (after the United States). In 1995 market shares of Taiwan's top four markets (United States, China, Japan, and Hong Kong) in Taiwan's exports were 24.0 percent, 15.3 percent, 10.6 percent, and 5.0 percent, respectively (excluding re-exports). Taiwan's exports to the United States have declined in absolute terms since 1987, and Hong Kong may soon become Taiwan's largest market if one includes Taiwan's exports via Hong Kong to China and also to third countries.

Table 2-8 shows China's direct and indirect trade with Taiwan. The data are obtained from table 2-7, with adjustment for the difference in f.o.b. and c.i.f. prices. That difference can be substantial because of complications in the shipping arrangements as well as the re-export margin earned by Hong Kong traders. By 1991 Taiwan had surpassed

Table 2-8. *China's "Direct" and Indirect Trade with Taiwan, 1986–95*
Millions of U.S. dollars; numbers in parentheses are percent shares of China's total imports or exports

	Imports			*Exports*		
Year	*Direct*	*Indirect*	*Total*	*Direct*	*Indirect*	*Total*
1986	20.2	846	866	0	126	126
	(0.05)	(2.0)	(2.0)	(0.0)	(0.4)	(0.4)
1987	81.0	1,147	1,228	0	257	257
	(0.2)	(2.7)	(2.8)	(0.0)	(0.7)	(0.7)
1988	208	2,357	2,564	0	418	418
	(0.4)	(4.3)	(4.6)	(0.0)	(0.9)	(0.9)
1989	688	3,048	3,746	93	513	606
	(1.2)	(5.2)	(6.3)	(0.2)	(1.0)	(1.2)
1990	1,342	3,450	4,792	320	670	990
	(2.5)	(6.5)	(9.0)	(0.5)	(1.2)	(1.6)
1991	2,991	4,889	7,880	595	989	1,584
	(4.7)	(7.7)	(12.4)	(0.8)	(1.4)	(2.2)
1992	4,140	6,611	10,751	698	987	1,685
	(5.1)	(8.2)	(13.3)	(0.8)	(1.2)	(2.0)
1993	6,136	7,915	14,051	1,063	966	2,029
	(5.9)	(7.6)	(13.5)	(1.2)	(1.1)	(2.2)
1994	7,219	8,971	16,190	1,468	1,134	2,602
	(6.2)	(7.8)	(14.0)	(1.2)	(0.9)	(2.2)
1995	9,223	10,406	19,629	1,936	1,381	3,317
	(7.0)	(7.9)	(14.9)	(1.3)	(0.9)	(2.2)

Sources: Estimated from table 2–7 by adjusting for c.i.f. (cost, insurance, and freight) and f.o.b. (free on board) price differences:

China's direct imports = 1.1 × Taiwan's direct exports
China's indirect imports = 1.2 × Taiwan's indirect exports
China's direct exports = Taiwan's direct imports/1.1
China's indirect exports = Taiwan's indirect imports/1.2.

Hong Kong and the United States to become the mainland's second largest supplier (after Japan). In 1995 the shares of Japan and Taiwan in China's imports were 22 percent and 15 percent, respectively. Taiwan's imports from the mainland are much smaller, but the rate of growth has been very high. In 1995 mainland products accounted for 3.7 percent of Taiwan's imports, and the mainland was the fifth largest supplier of Taiwan after Japan (29.2 percent), the United States (20.1 percent), Germany (5.5 percent), and South Korea (4.2 percent). Taiwan also became a significant market for the mainland, accounting for 2.2 percent of China's exports (table 2-8).

Trade and Investment between Hong Kong and Taiwan

Before 1987 economic ties between Hong Kong and Taiwan were one-sided because of Taiwan's trade protectionism and foreign exchange controls. In the mid-1970s Hong Kong became Taiwan's third largest market (after the United States and Japan), accounting for roughly 7 percent of Taiwan's exports. However, barriers against Hong Kong goods in Taiwan were high. Hong Kong was the third largest investor in Taiwan after the United States and Japan, but Taiwanese investment in Hong Kong was insignificant because of Taiwan's then stringent foreign exchange controls. By the end of 1989 Hong Kong investment in Taiwan totaled $1.2 billion (11 percent of total inward investment in Taiwan), while U.S. and Japanese investments were $3 billion and $2.9 billion, respectively. By the end of 1991 Hong Kong's cumulative investment in Taiwan totaled $1.6 billion.[23]

Economic ties developed rapidly in the late 1980s, however, with the liberalization of Taiwan's import and foreign exchange controls, the sharp appreciation of the Taiwanese currency, and Taiwan's use of Hong Kong as an intermediary in its interactions with China. Many Taiwanese toured Hong Kong on their way to the mainland, with Taiwan becoming the foremost source of tourists for Hong Kong from 1988 to 1993 (20 percent of tourist arrivals in 1993).[24] Taiwan also became a significant investor in Hong Kong. The share of the Taiwan market in Hong Kong's domestic exports jumped from 1.0 percent in 1985 to 3.4 percent in 1995, amounting to $1,031 million. Since 1986 Taiwan has been the seventh largest market for Hong Kong (after China, the United States, Singapore, Germany, Japan, and the United Kingdom).

In the 1990s Hong Kong was the fourth market for Taiwan after the United States, China, and Japan. Taiwan's 1995 exports retained for internal use in Hong Kong were $6.3 billion, or 5.7 percent of Taiwan's exports (table 2-6). After adjustment for the cost of insurance and freight, Hong Kong's 1995 retained imports from Taiwan were $6.6 billion (8.3 percent of the total retained imports of Hong Kong). Taiwan is the fourth largest supplier of Hong Kong's retained imports, after Japan, the United States, and China.

Taiwanese investment in Hong Kong also soared. Cumulative investment from Taiwan reached $2 billion by the end of 1989, and half of the amount was invested after 1987.[25] Taiwan became the fifth largest investor in Hong Kong after the United Kingdom, the mainland, Japan, and the United States. By the end of 1995 cumulative investment from Taiwan was estimated to be $4 billion to $5 billion.[26] Services accounted for over half the investment, and export-import trade accounted for over 20 percent. The rest was mainly in finance and insurance.

Integration of the Labor Market in the China Circle

The labor market lacks integration among the trio. Controls against visitors from the mainland entering both Taiwan and Hong Kong are very strict, especially in Taiwan's case, though illegal migrants from the mainland are common. China and Hong Kong agreed in 1982 to a quota restricting the number of migrants from China to 75 a day (27,375 a year). Although the quota was increased to 105 a day in 1993 and further increased to 150 a day in 1995, mainland relatives of Hong Kong residents still wait for a long time before they can migrate to Hong Kong. Indeed, mainland spouses of Hong Kong residents usually have to wait ten years before they can migrate to Hong Kong.

Although Hong Kong barriers against permanent migration from the mainland are very high, the barrier against temporary stays was lowered in recent years. Since 1989, when the Hong Kong government embarked on its first labor importation scheme, increasingly more Chinese workers came to Hong Kong to work on temporary contracts. The third labor importation scheme (1992) further doubled the labor intake quota to 25,000, practically all filled by Chinese workers. But the quota was reduced to 2,000 at the end of 1995 bcause of a rise in unemployment in Hong Kong. Barriers against professionals from China were also relaxed recently. Since 1990 Hong Kong employers have been permitted to employ mainland professionals who have stayed overseas for more than two years. In April 1994 the Hong Kong government announced a trial scheme to import 1,000 mainland graduates.

There are few restrictions against Chinese entering Hong Kong on official passports, and it is estimated that, by March 1994, 65,600 cadres were working in PRC companies in Hong Kong.[27] A substantial number of illegal immigrants and short-term visitors also participate illegally in the labor market. In only a few years tourists from the mainland increased from a trickle to more than 2.2 million in 1995, and China surpassed Taiwan as a source of tourists for Hong Kong in 1994. China has few controls against Hong Kong and Taiwan residents working on the mainland. In October 1995, 122,000 Hong Kong residents reported that they had worked in China during the previous year.[28] But the Basic Law, or the future constitution of Hong Kong after its reversion to China, stipulates that direct relatives of Hong Kong residents have the right to enter Hong Kong. In 1994 there were 75,000 direct relatives (children or spouse) of Hong Kong residents living in China. This number will grow as Hong Kong–mainland marriages become increasingly common. Strict controls against immigration from the mainland must be relaxed in 1997.

Problems and Prospects of the China Circle

Although the growth of the China Circle has been extremely rapid, and many projections of the future are rosy, the China Circle faces a host of problems, both economic and political. Four problems are discussed here: the difficulty of further reforms of the Chinese economic system; problems of adjustment to China's export drive in the world market, especially trade frictions with the United States and other OECD countries; maintaining prosperity in Hong Kong after July 1, 1997; and tensions across the Taiwan Strait.

Difficulty of Further Reforms

After eighteen years of economic reforms and opening, the Chinese reforms have reached a crossroads. Most of the early reforms (agricultural reforms, privatization of small firms) have been completed, and the remaining reforms (reform of large state-owned enterprises, banking reforms, reforms of the fiscal and monetary system) will be much more difficult. The lack of thorough reforms will limit foreign investment in China, especially in industries selling to the domestic market, and also in service sectors such as banking, finance, and telecommunications. The investment of Hong Kong and Taiwan in China has been concentrated in export sectors, a fact that can be attributed to the lack of thoroughness of China's reforms. China merely provides cheap land and labor; all the other ingredients such as raw materials, management, industrial design, marketing, and financing are provided externally. Moreover, the foreign investor can recoup the investment as the product is exported, earning foreign exchange in return. But because of China's foreign exchange controls, foreign investors who invest in industries selling to the domestic market find it more difficult to recoup their investment. Investors in sectors such as banking, finance, and telecommunications will face a hard time, since such sectors tend to be highly regulated. The speed of opening these sectors to foreign investors will be limited by the pace of China's reforms.

Problems of Adjustment in the World Market

Although the economies of the China Circle are tightly linked, it must be stressed that the China Circle is not an inward-looking trade bloc. The economic reality of the Circle is that the United States is its largest market and Japan is its largest supplier of capital goods and technology. An inward-looking trade bloc excluding Japan and the United States would be detrimental to the Circle. Nevertheless, the scale of China's entry to the world market has raised problems of adjustment in the world economy. China's exports are overconcentrated on the sale of labor-intensive products to the U.S. market. During several months in 1996 the U.S.-China bilateral deficit surpassed the U.S.-Japan deficit to become the largest bilateral deficit in U.S. trade. Though the size of bilateral trade deficits has little economic significance, it can easily lead to severe political problems and trade frictions.

Protectionism is a problem for China. China's most-favored-nation (MFN) status in the United States is subject to renewal debates every year. China is running up against its quota of textiles and clothing exports, and the number of antidumping charges against China is increasing very fast. Almost every year China is high on the hit list of U.S. section 301 or super 301 investigations. In 1996 China's exports increased only 3 percent, after growing at double-digit rates for more than a decade. Although the poor performance appears to be mainly a result of the failure of China to give speedy rebates of value-added taxes for exports, protectionism may also play a role.

The trade problems have been exaggerated, however. Since the relocation of the export-oriented industries of Hong Kong and Taiwan to China, the trade deficit of the United States with China has increased tremendously, whereas the bilateral deficits of the United States with Hong Kong and Taiwan have declined. On the one hand, the total exports of the China Circle to the United States have gone up, as the relocated firms often expand in scale with the availability of cheap labor and land in China. On the other hand, the total exports of the United States to the China Circle have also increased markedly owing to Taiwan's trade liberalization and rapid growth in China. The net result is that the size of America's overall trade deficit with the China Circle grew from $22.6 billion in 1989 to $39.6 billion in 1995, but the size of the deficit relative to U.S. exports stayed roughly constant at 7 percent.

These trade frictions can be avoided if China liberalizes its imports of commodities and services. Because services can usually be provided only on site, the liberalization of service trade often implies the liberalization of foreign investment in service industries. That would give rise to many opportunities for investment and would also mean a much higher degree of integration for the China Circle.

Even if China liberalized its imports of goods and services, Hong Kong and Taiwan are unlikely to become the foremost suppliers of China. Though China may become the largest market for Hong Kong and Taiwan, the reverse will probably not occur. If China liberalizes its imports, China's imports of final goods will increase relative to its imports of semimanufactures, and China is likely to shift its imports from Hong Kong and Taiwan to Japan and the West. China must look outside the China Circle for its capital goods, technology, and market. The liberalization of China's imports will thus imply rich opportunities for East Asia and the world.

Stability of Hong Kong after 1997

While the economic freedoms of Hong Kong are spelled out in detail in the Sino-British declaration on the future of Hong Kong, it must be remembered that "one country, two systems" is an untried formula and that the future stability and prosperity of Hong Kong is not assured. Because Hong Kong is the pivot of the economic integration of the China Circle, the consequences for China and the China Circle can be very serious.

The workability of "one country, two systems" lies more in the political and social realms than in the economic realm. The political problem is that China must resist the great temptation and pressure to intervene in the internal affairs of Hong Kong. So long as China can respect the autonomy of Hong Kong, the economic fundamentals are likely to work in Hong Kong's favor. As I have elaborated elsewhere, the prospect of Hong Kong as China's middleman is bright.[29] China will further decentralize its trading system and create greater opportunities of intermediation for the Hong Kong middleman. Moreover, significant economies of scale and agglomeration in trading activity will be realized; it will also be difficult for other cities like Singapore or Shanghai to compete with Hong Kong, since Hong Kong is the established center for China's trade. The existence of economies of scale in intermediation will enhance the demand for the middleman, for small firms will not be able to trade efficiently.

Traders tend to agglomerate in a city, suggesting that important external economies are involved. Once a city acquires a comparative advantage in trade, the advantage feeds on itself; more trading firms will come to the city, making the city even more efficient in trade. At present investors seem to be bullish about Hong Kong, and they have propelled the prices of Hong Kong's real estate market and stock market to record heights. Emigration among Hong Kong's professionals has slowed down, while migrants are returning in increasing numbers.

Competitors of Hong Kong. The competitors to Hong Kong's regional role are Singapore, Taipei, and Shanghai. Of the three cities, Singapore is the toughest competitor, since its quality of skills, facilities, and infrastructure rival that of Hong Kong. However, Singapore and Hong Kong are more complementary than competitive because they serve different regions. Hong Kong and Singapore are separated by a four-hour flight. Singapore is ideally situated to serve Southeast Asia, whereas Hong Kong is better located to serve China, Taiwan, Korea, and even Japan. Though both Southeast and Northeast Asia are dynamic, the size of the Northeast Asian economy is much larger than the Southeast Asian economy; this gives Hong Kong an edge over Singapore as a regional service center. Moreover, Hong Kong has a greater range of skills, for the size of the Hong Kong economy is 2.5 times that of Singapore. Because of its location Singapore cannot seriously challenge the lead of Hong Kong as the gateway to China. In fact, Singapore has become increasingly dependent on Hong Kong for its trade with China.[30]

Taipei, which aspires to become a regional financial, corporate, and transportation center, can pose a serious threat to Hong Kong because of proximity. Taipei is located in the same region as Hong Kong and is only one hour's flight from Hong Kong. Moreover, Taiwan has huge investments and a big trading network throughout Southeast Asia. Although Taiwan's investment in China is only a fraction of Hong Kong's, it has grown very fast and may rival Hong Kong's investment in the long run. Finally, Taiwanese speak better Mandarin than Hong Kong Chinese, and Taiwanese businessmen are more adept at dealing with bureaucrats because of a similar business environment in Taiwan.

Currently Taipei fails to realize its potential because of the official policy of no direct links with the mainland. As discussed, Taiwan's trade and investment with the mainland are largely handled through Hong Kong, and Taiwan's policy strengthens Hong Kong's lead as a regional service center. Taipei can only be a serious competitor to Hong Kong in the very long run, when direct Taiwan-China links are fully developed. Even then, Taipei will still face challenges from Hong Kong. Taipei's service skills lag far behind those of Hong Kong, and Taiwan's heavily regulated environment is not conducive to the development of Taipei as a regional service center. Furthermore, Hong Kong's natural hinterland, Guangdong, is much more dynamic than Taiwan's natural hinterland, Fujian. Finally, it must be remembered that Hong Kong and Taiwan are in some ways complementary. Taiwan is stronger in engineering and industrial skills, whereas Hong Kong is stronger in finance, shipping, and legal services.

Shanghai has the advantage of location; it is situated at the mouth of the Yangtze basin, the largest and most prosperous river basin in China. Though Shanghai is set to become the domestic financial center of China, it cannot compete with Hong Kong as a regional or international center. Shanghai cannot become a serious contender for being a regional or international financial center unless the renminbi can achieve convertibility on the capital account. Even though China achieved convertibility in merchandise trade in 1996, capital account convertibility will take much longer. The experiences of trade and exchange rate liberalization in developing countries have repeatedly demonstrated that capital account convertibility takes a long time to achieve.

The development of infrastructure and service skills is a time-consuming and capital-intensive process. Moreover, the development of services is dependent on the regulatory environment. Given the corrupt, inefficient, and immense bureaucracy in China, developing an efficient international service center will be very difficult. Finally, Shanghai does not have a good port because of the silting of the Yangtze river. Ships over 50,000 tons can enter Shanghai (or any port along the Yangtze) only at high tide. Though Shanghai is emerging as the business center of the Yangtze Delta, it cannot challenge the position of Hong Kong as a regional and international business and financial center.

Prospect of Hong Kong as a Trading Center and Service Hub. There has been concern that the outward-oriented labor-intensive processing operations of the China Circle do not have a robust future, because of rising world protectionism and increasing competition from Vietnam, other ASEAN nations, and other coastal provinces in China. The slowdown of China's export growth to only 1.5 percent in 1996 seems to confirm those fears. There is also speculation that Hong Kong may be losing its status as China's premier trading center as the focus of China's development effort shifts from Guangdong to Shanghai. Such concerns are premature. First, the 1996 stagnation in China's exports appears to be temporary. China slashed the rebate rate of value-added tax on exports from 17 percent in 1994 to 8.9 percent on January 1, 1996. This led to a rush to export in 1995 and a corresponding slowdown in 1996. Second, Guangdong's exports growth has remained high. Though Guangdong's share of China's exports fell from the 1994 record of 41.5 percent to 38 percent in 1995, Guangdong's share rebounded to 39.7 percent in 1996. Guangdong's export dynamism appears to remain strong.

Finally, despite the stagnation in China's exports in 1996, China's trade with Hong Kong has continued to grow, and Hong Kong's share of China's trade has increased. Table 2-9 shows the share of China's trade through Hong Kong and through Shanghai since 1979. The share of China's trade (imports plus exports) through Hong Kong rose rap-

Table 2-9. *China's Trade via Shanghai and Hong Kong, Selected Years, 1979–96*

Millions of U.S. dollars; numbers in parentheses are percent shares of China's total trade

	Via Shanghai[a]			*Via Hong Kong with third countries*[b]			*With Hong Kong*[c]		
Year	*Total*	*Exports*	*Imports*	*Total*	*Exports*	*Imports*	*Total*	*Exports*	*Imports*
1979	n.a.	3,675	n.a.	1,224	945	279	3,278	2,872	406
	n.a.	(26.9)	n.a.	(4.2)	(6.9)	(–1.8)	(11.2)	(21.0)	(2.6)
1985	14,873	4,908	9,965	9,972	3,711	6,261	15,355	7,027	8,328
	(21.4)	(18.0)	(23.6)	(14.3)	(13.6)	(14.8)	(22.1)	(25.7)	(19.7)
1987	15,635	6,601	9,034	17,200	9,021	8,179	25,906	13,939	11,967
	(18.9)	(16.7)	(20.9)	(20.8)	(22.9)	(18.9)	(31.3)	(35.3)	(27.7)
1989	19,437	7,711	11,726	34,210	20,146	14,064	42,994	23,049	19,945
	(17.4)	(14.7)	(19.8)	(30.6)	(38.3)	(23.8)	(38.5)	(43.9)	(33.7)
1990	17,289	8,662	8,627	39,137	24,065	15,072	49,379	27,856	21,523
	(15.0)	(14.0)	(16.2)	(33.9)	(38.8)	(28.3)	(42.8)	(44.9)	(40.3)
1991	20,409	10,151	10,258	51,292	30,457	20,835	62,862	34,633	28,229
	(15.1)	(14.1)	(16.1)	(37.8)	(42.4)	(32.7)	(46.3)	(48.2)	(44.3)
1992	25,145	11,964	13,181	66,768	37,943	28,825	78,929	41,685	37,244
	(15.2)	(14.1)	(16.4)	(40.3)	(44.7)	(35.8)	(47.7)	(49.1)	(46.2)
1993	30,931	13,977	16,954	80,596	42,976	37,620	93,629	47,325	46,304
	(15.8)	(15.2)	(16.3)	(41.2)	(46.8)	(36.2)	(47.8)	(51.6)	(44.5)
1994	36,246	18,938	17,308	94,326	50,044	44,282	108,029	55,379	52,650
	(15.3)	(15.7)	(15.0)	(39.9)	(41.4)	(38.3)	(45.7)	(45.8)	(45.5)
1995	48,138	25,608	22,530	110,972	58,309	52,663	124,868	63,490	61,378
	(17.1)	(17.2)	(17.1)	(39.5)	(39.2)	(39.9)	(44.5)	(42.7)	(46.5)
1996	52,869	27,213	25,656	119,850	62,594	57,256	132,821	67,120	65,701
	(18.2)	(18.0)	(18.5)	(41.3)	(41.5)	(41.3)	(45.8)	(44.5)	(47.3)

Source: Trade via Shanghai: data since 1993 from *China Customs Statistics*; data before 1993 from *The Foreign Economic Statistical Yearbook of Shanghai* (Shanghai: Statistical Bureau Shanghai [internal document], 1994); data for China's trade via Hong Kong and with Hong Kong from: *Hong Kong External Trade*, various issues (adjusted for f.o.b. and c.i.f. price differences).

n.a. Not available.

a. Includes products imported or exported by other provinces via Shanghai's customs.

b. Includes only China's trade with third countries in the form of Hong Kong re-exports.

c. Includes also China's exports retained in Hong Kong and China's imports of Hong Kong goods in addition to China's trade via Hong Kong.

idly from 4.2 percent in 1979 to a peak of 41 percent in 1993, and has since leveled off, amounting to 41.5 percent in 1996. This is in sharp contrast to the rapid decline of Shanghai's share through 1990 and the sharp increase since. In 1996 Hong Kong's share was still more than twice the 18 percent share of Shanghai. Moreover, transshipment and direct shipments handled by Hong Kong firms are not included in the trade figures shown here. As discussed earlier, in recent years Hong Kong traders have substituted transshipment and direct shipments for re-exports. The slight fall in the share of China's trade in the form of Hong Kong re-exports from 1993 to 1995 does not indicate the decline of Hong Kong's middleman role in China's trade.

It is amazing that seventeen years after China's opening, Hong Kong's shares are still so high. In 1995 Hong Kong still handled 40 percent of China's trade in the form of re-exports (table 2-9), and Hong Kong's share would be two-thirds if transshipments, direct shipments, and trade with Hong Kong itself were included. Hong Kong also accounted for 46 percent of the 1995 contracted FDI in China.

The continuing high shares of Hong Kong in China's trade and investment show the importance of economies of scale and agglomeration in trading and service activities. As business activities agglomerate, land prices and wages rise, putting a stop to the tendency toward agglomeration. Agriculture is the first activity to move out of the city because it is land intensive. Manufacturing is the second activity to move, and many Hong Kong manufacturing firms are moving their land-intensive and labor-intensive processes to China. Transportation is the next activity to be affected, since ports and airports are also land intensive. The high costs of Hong Kong's new airport and container ports are expected in land-scarce Hong Kong. The new Hong Kong airport will be twenty times as expensive as the new Huangtian airport in Shenzhen, which will also have two runways when it is fully developed. Were it not for political barriers, it would be more rational to put Hong Kong's new airport in Shenzhen. However, China's customs administration is not noted for its efficiency: it is too risky for Hong Kong to put its new airport in Shenzhen unless China thoroughly reforms and modernizes that administration. Such reforms cannot realistically be expected in the near future. But when such reforms are completed, it will be rational for Hong Kong to divert some of its cargo traffic to China.

It is natural that Hong Kong firms have recently substituted direct shipment for re-exports through Hong Kong in handling China's trade. Transportation via Hong Kong can be expensive because of the congested container ports and airport. Congestion also generates huge external costs for the residents of Hong Kong. To forever allow nearly half of China's rapidly increasing trade to be transported via Hong Kong's tiny territory is not in Hong Kong's best interest.[31] Trading, business services, and financial services will be least affected by rising land prices and wages, because such activities are neither land intensive nor labor intensive. New York and London long ago lost their comparative advantage in manufacturing, but their positions in trading, business, and finance remain formidable.

The Chinese are establishing many trading companies in Hong Kong, showing that they recognize the territory's efficiency in trading. Some Hong Kong traders fear competition from Chinese trading companies in Hong Kong. The situation is not a zero-sum game, however; because of economies of agglomeration the arrival of Chinese trading companies further enhances the position of Hong Kong as a trading center. Hong Kong's shares in China's trade and investment are so high that they are unlikely to rise further. Rather, the shares are likely to decline, for China is building many ports and many foreign multinationals are investing in China. In the long run China will probably overcome its transportation bottlenecks and acquire modern trading skills. Even if China cleans up its bureaucracy, it will still rely on Hong Kong for trade, financial, and business services because of economies of scale and agglomeration.

Tensions across the Taiwan Strait

China's military exercises in the Taiwan Strait in late 1995 and early 1996 reminded the world that the continuous economic integration of the China Circle is not a forgone conclusion.[32] After the reelection of Lee Tenghui to the presidency in Taiwan, Beijing conducted a high-level meeting on the Taiwan issue in May 1996. Beijing resumed its policy of wooing Taiwanese businessmen. Taiwan has not reciprocated the recent friendly gestures from Beijing. On August 15, 1996, Lee Tenghui announced that there should be restrictions on Taiwanese companies that want to invest in the mainland.[33] Though the precise nature of the restrictions has not been clarified, businessmen have reacted unfavorably to the announcement. Formosa Plastics suspended a $3 billion power plant project in China, and food giant President Enterprises halted its $100 million project to set up a power plant in the central city of Wuhan.

Moreover, Taiwan's plan to develop Taipei as a regional operations center has suffered a serious setback; investors are nervous about a possible outbreak of hostilities in the future. On August 20, 1995, China announced regulations on direct shipping with Taiwan, allowing Chinese, Taiwan, and even foreign shipping companies to sail directly between the sides.[34] Since the announcement, many nominally foreign shipping companies owned by Taiwanese have applied for permission to start direct shipping services, and apparently Beijing will soon grant formal approvals. But according to Taiwanese rules promulgated in May 1995, only foreign-registered vessels are allowed to sail directly between an exclusive tariff-free zone near Kaohsiung and Chinese ports. Such vessels are allowed to ship only non-Taiwanese goods to China. Because the tariff-free zone is outside Taiwanese customs, it is not considered to be a violation of Taiwan's general prohibition on direct shipping. The zone was formed to promote transshipment via Kaohsiung. Once Beijing approves foreign shipping companies to operate across the Taiwan Strait, foreign-registered vessels can start to carry cargo between the two coastal ports in Fujian (Xiamen and Fuzhou) and third countries via Kaohsiung. This is a far cry from full-scale direct shipping, and the impact on Hong Kong is estimated to be less than 1 percent of Hong Kong's shipping. The development of direct shipping links across the Taiwan Strait will probably be slow because of mutual suspicion. The Taiwanese government's policy is to delay direct links for as long as possible, since it believes that the prohibition on direct links is one of its last bargaining chips.

Prospects of the China Circle

Economic forces point to a rapid continuation of the economic integration of the China Circle. Though political uncertainties exist, such as the stability of China in the post-Deng era, doubts on the viability of Hong Kong after 1997, and possible hostilities over the Taiwan Strait, the economic fundamentals of the China Circle are strong.

Because of the many differences in the political, legal, and economic systems between the mainland on the one hand, and capitalist China (Hong Kong and Taiwan) on the other, the economic integration of the mainland with capitalist China will be highly uneven. Integration will proceed rapidly in some areas but slowly in others. Between the mainland and capitalist China, controls on movements of goods are relatively liberal, whereas controls on capital and foreign exchange are strict and controls on migration are even stricter. The integration of the commodity market between the mainland and capitalist China will proceed rapidly owing to the relatively mild controls on the flow of goods. However, even for the commodity market, one should distinguish between export-processing industries and import-competing industries. The outward processing operations of capitalist China on the mainland have developed extremely rapidly because its products are exported and it is not hampered by China's foreign exchange controls. The growth of external investment in China's import-competing industries will necessarily be slower because of China's foreign exchange controls.

The integration of services industries between the mainland and capitalist China will similarly be slow, since most services cannot be exported and are sold in the domestic market. Moreover, services are performed on people and require people-to-people contacts. The controls of capitalist China on migration from the mainland will hamper the full integration of services. The integration of the financial markets between the mainland and capitalist China will also be quite slow, for China's foreign exchange controls on the capital account will most likely be strict even in the medium term. And the integration of the labor markets between the mainland and capitalist China will probably be very slow because of controls on migrations.

Notes

1. Macao's economy is much smaller than Hong Kong's and can be regarded as an appendage to the Hong Kong economy. This chapter concentrates on economic interdependence among China, Hong Kong, and Taiwan—the "trio"—and there will be no separate treatment of Macao.

2. Whereas the bilateral economic relations between Hong Kong and Guangdong, Taiwan and Fujian, and Hong Kong and Taiwan are quite close, the economic ties between Guangdong and Fujian are not particularly strong. The two provinces lack complementarity, since both lack natural resources and are at the same stage of economic development. Nor are the two provinces highly rivalrous. They are linked to separate communities of overseas Chinese with different dialects, which moderates their competition for overseas investment.

3. Chia Siow Yue, "Motivating Forces in Subregional Economic Zones," Occasional Papers (Honolulu: Pacific Forum/Center for Strategic and International Studies, December 1993).

4. Robert A. Scalapino, "The United States and Asia: Future Prospects," *Foreign Affairs,* vol. 70 (Winter 1991–92), pp. 19–40.

5. Chia, "Motivating Forces in Subregional Economic Zones."

6. *Hong Kong Economic Journal,* May 5, 1994.

7. Sung Yun-Wing, "Non-Institutional Economic Integration via Cultural Affinity: The Case of Mainland China, Taiwan, and Hong Kong," Occasional Paper 13 (Chinese University of Hong Kong, Hong Kong Institute of Asia-Pacific Studies, July 1992), p. 8.

8. Milton Yeh, "Ask a Tiger for Its Hide? Taiwan's Approach to Economic Transaction across the Straits," in Jane Khanna, ed., *Southern China, Hong Kong, and Taiwan* (Washington: Center for Strategic and International Studies, 1995), pp. 61–70.

9. *Asian Times,* August 8, 1996.

10. Hong Kong Census and Statistics Department, *External Investments in Hong Kong's Non-Manufacturing Sectors, 1993 and 1994* (Hong Kong 1996).

11. Sung Yun-Wing, "Chinese Outward Investment in Hong Kong: Trends, Prospects, and Policy Implications," Technical Paper 113 (Paris: OECD Development Center, July 1996), p. 17.

12. The "round tripping" of Chinese capital in Hong Kong (Chinese capital that flows to Hong Kong and then back to China to be recorded as foreign investment so as to capture the benefits given to foreign investors) inflates both the figures of Chinese investment in Hong Kong and Hong Kong investment in China.

13. As a result of pressure from the United States, China tried to trace the final destination of its exports via Hong Kong, and a substantial portion of its exports to Hong Kong were reclassified as exports to final destinations in 1993. Because of this reclassification, China's exports to Hong Kong dropped by 41 percent, and China's exports to the United States, Japan, and Germany grew by 97 percent, 35 percent, and 62 percent, respectively. Despite the reclassification, a substantial portion of China's exports via Hong Kong to third countries is still classified as exports to Hong Kong because China is unable to trace the final destination of all its exports via Hong Kong.

14. Such adjustments were first made in the monthly newsletter of the Hong Kong Bank. See *Economic Report of the Hong Kong Bank,* January 1991.

15. This is Guangdong's share of Hong Kong's imports involving outward processing from China.

16. Hong Kong Trade Development Council, "Hong Kong's Trade and Trade Supporting Services," April 1996, pp. 8–9.

17. It is assumed that China's share of transshipments and direct shipments of third-country products handled by Hong Kong traders is the same as China's share of Hong Kong's re-exports. China has been the foremost market of Hong Kong's re-exports, with a share of 34.5 percent in 1995.

18. In 1994 a group of Taiwanese tourists were murdered in a ship at the Thousand Island resort in Zhejiang. The connivance of local authorities was suspected and there was widespread protest in Taiwan over China's handling of the case.

19. Kao Chang and Sung Yun-Wing, *An Empirical Study of Indirect Trade between Taiwan and Mainland China* (Taipei: Mainland Committee of Taiwan Government, September 1995), p. 79.

20. Part of the "missing exports" represent Taiwan's transshipment through Hong Kong to third countries rather than to China. The fraction of 80 percent is used by Taiwan's Ministry of Trade to estimate Taiwan's exports to the mainland.

21. Sung Yun-Wing and others, *The Fifth Dragon: The Emergence of the Pearl River Delta* (Singapore: Addison-Wesley, 1995), p. 77.

22. Since 1993 China's statistics have vastly overstated China's direct exports to Taiwan because of the 1993 reclassification that tried to include China's indirect exports via Hong Kong. See note 13.

23. Hong Kong Trade Development Council, "Hong Kong's Economic Relationship with Taiwan," February 1992, p. 4.

24. The mainland has replaced Taiwan as the foremost source of tourists for Hong Kong since 1994, partly owing to the surge of visitors from the mainland to Hong Kong, and partly owing to the drop in Taiwanese tourists as a result of the Thousand Island incident and hostilities over the Taiwan Strait in 1995 and 1996.

25. Zhou Ba Jun, *Hong Kong: The Economic Transition accompanying the Political Transition* (in Chinese) (Hong Kong: Joint Publishing Co., 1992), p. 167.

26. *Hong Kong Daily News,* August 8, 1996.

27. Lin Tzong-biau and Kan Chak-yuen, "Chinese Firms and the Hong Kong Economy" (in Chinese), *Asian Studies* (Chu Hai College, Center for Asian Studies, Hong Kong), no. 17 (April 1996), pp. 130–76.

28. *General Household Survey* (Hong Kong Government, Census and Statistics Department, 1997).

29. Sung Yun-Wing, *The China–Hong Kong Connection: The Key to China's Open Door Policy* (Cambridge University Press, 1991), pp. 28–42.

30. Sung, *China–Hong Kong Connection,* pp. 135–36.

31. Moreover, the average gross margins earned by Hong Kong's trading firms (16.1 percent for exporting Chinese goods and 12.6 percent for exporting third-country goods) appear to be the same for re-exports through Hong Kong and for offshore trade. See Hong Kong Trade Development Council, "Hong Kong's Trade and Trade Supporting Services," pp. 8–9.

32. However, the episode also demonstrated that China's naval power is not yet strong enough to carry out an invasion.

33. *Hong Kong Economic Journal,* August 16, 1996.

34. *South China Morning Post,* August 21, 1996.

CHAPTER THREE

Economic Policy Reform in the PRC and Taiwan

Barry Naughton

FOR MOST of the period after 1949, the People's Republic of China and Taiwan were officially at war. Tensions eased during the 1980s, but the two sides never signed any formal agreements, because neither side was willing to recognize the legitimate existence of the other. Despite these obstacles, by the early 1990s the two were among each other's largest trading partners, and Taiwan had become the second largest investor in the mainland. Yet in 1995 relations took another sudden turn for the worse, and on three occasions in 1995 and 1996 the PRC military lobbed ballistic missiles into the ocean near Taiwan's largest seaports. Despite the tensions, trade and investment continued to grow in 1995–96, but the tensions and rivalry in the relationship, which had been submerged while economic ties developed, suddenly resurfaced for all to see.

Obviously, economic cooperation between these two distant neighbors is not governed by state-to-state agreements. Instead, economic integration has advanced without any international framework whatsoever. Yet governments on both sides had to modify preexisting policies to facilitate the growth of economic ties. This chapter examines the process and sequence by which enabling policy reforms were adopted on the two sides of the Taiwan Strait. Policy changes are interpreted primarily as responses to changing economic conditions, but given the political rivalry between the PRC and Taiwan political factors are never completely absent.[1] Indeed, the rivalry between the two shapes the way that economic benefits are perceived and weighed by the two sides.

The framework yields a number of surprising insights. It explains why 1987 was the crucial year in which the economic relationship between China and Taiwan blossomed, and it throws some light on the causes of the sudden deterioration of relations in 1995. Moreover, a careful look at the evolution of economic policy in the PRC and in Taiwan reveals some surprising patterns on both sides. Of the PRC it is usually held that economic reforms proceeded tentatively and experimentally, without a prior blueprint. The Chinese expression "crossing the river by groping for stepping stones" is often used to describe the overall reform process.[2] Conversely, the government of Taiwan is often held up as a "hard state," possessing substantial economic expertise and able to shape Taiwan's development according to its design.[3] It is therefore surprising to find that the policy toward each other developed exactly contrary to these generalizations. PRC policy toward Taiwan trade and investment was overwhelming pro-active. There is substantial evidence that after 1985 Chinese Premier Zhao Ziyang conceived and implemented a coordinated set of policies designed to achieve exactly what was in fact achieved: the attraction of a large volume of Taiwan investment to coastal China and the integration of parts of coastal China into preexisting export networks created by Hong Kong and Taiwanese businesses. In contrast, Taiwan's policy was largely reactive. Adjustment to economic success was delayed, and response to the mainland was belated.

A basic reason for that difference between Taiwan and the PRC was the fundamental asymmetry in their respective positions. The PRC is huge and politically important, enjoying diplomatic recognition from more than one hundred countries. Taiwan is small and is virtually invisible on the international diplomatic scene. Conversely, Taiwan is fairly rich and has impressive commercial and technological capabilities, while China is still relatively poor and inexperienced in the global marketplace. Thus Taiwan seeks to translate its economic assets into political recognition, while the PRC seeks to prevent it from doing so. In this contention, the central issue is the legitimacy of the Taiwan government as a sovereign state. For many years, the governments in Beijing and Taipei agreed that there could only be one legitimate government of China but disagreed violently over which of them was that government.[4] In 1988, a few months after he succeeded Chiang Ching-kuo as president, Lee Teng-hui began to relax this position and agreed to participate in multilateral organizations such as the Asian Development Bank and the Asia-Pacific Economic Cooperation (APEC) group alongside the PRC. From that point on, Taiwan accepted the existence of the Beijing government and the fact of two competing governments with steadily increasing explicitness, without relaxing its own insistence that it be treated as a legitimate and sovereign government. For its part, the PRC accepted cooperation in regional institutions precisely because Taiwan agreed to participate without all the trappings of full sovereignty, agreeing to enter as "Chinese Taipei." A common ground was found without in any sense resolving the fierce disagreement over whether Taiwan was a sovereign national government. Subsequently, acute conflict is always possible when Taiwan is granted symbols of the legitimacy accorded a sovereign government.

From Beijing's standpoint any cooperation between the PRC and Taiwan in which Taiwan is not granted the trappings of a sovereign state is a desirable one, because it implicitly lends support to Beijing's claim that Taiwan's government is merely that of a province of China. Economic contacts, since they take place between businesses, not governments, can always be construed as nonofficial. Thus Beijing can pursue economic benefits while also arguing that it is "undermining" the official status of the Taiwanese government. From Taipei's standpoint, cooperation with the PRC or participation in international organizations is potentially desirable: it raises Taiwan's political profile and provides a forum in which economic influence could be traded for political standing. However, Taipei must balance the desire to participate against the perceived indignity of being forced to participate on the basis of less than full sovereignty. The government in Taipei therefore has mixed and complex motives. On the one hand, economic considerations dictate seizing the opportunities presented on the China mainland; and in Taiwan's increasingly democratic political system, the individuals that benefit from such opportunities increasingly form interest groups that push for specific policy measures. On the other hand, the government of Taiwan has a long-term interest in delaying economic contacts and holding them hostage to a greater degree of official recognition.

These political calculations have been relatively constant in the evolving relationship between the PRC and Taiwan. But as economic factors have changed, the mix of economic and political factors has shifted, lending an apparent volatility to the relationship. In fact, the basic objectives of the two sides have not changed much over time. As the relationship has shifted from a distant to a close one, the struggle for incremental advantage has become less overt, but no less intense. Both sides know how to drive a hard bargain, and both sides know the stakes are high. These considerations hem the two sides into a fairly narrow framework as they grope toward greater economic cooperation.

Prelude to Cooperation

Immediately upon China's shift to a policy of economic reform, it launched an energetic campaign to appeal to Taiwan. On January 1, 1979, Beijing stopped the perfunctory shelling of the Taiwan-held islands off the Fujian coast that had been going on since the 1950s (explosive shells had been replaced with leaflets in the early 1970s, after President Nixon's trip to China). The National People's Congress issued a "Letter to Taiwan Compatriots," which called for direct mail and shipping and trade relations, while top leaders declared their commitment to peaceful reunification. The Chinese Ministry of Foreign Trade established a working group in Hong Kong that dealt secretly with Taiwanese business people and made large purchases of Taiwanese goods.[5] In 1980 Beijing made the astonishing decision that Taiwanese manufactured products could enter China duty-free, since they were the products of a Chinese province. Needless to say, Taiwanese businesses took advantage of these provisions, and Taiwan's exports to China surged from $21 million in 1979 to $390 million in 1980.*

These initiatives, however, did not evoke a serious response from Taiwan. The Taiwanese government responded with its policy of "three nos": no contacts, no negotiation, no compromise. Taiwan had little or no economic incentive to become involved with the mainland at that time. Exports to the U.S. market were growing rapidly, and that was still the focus of Taiwan's economic policy. Moreover, the mainland's economic reform was still in its infancy, and a minimum essential degree of marketization had probably not been achieved. In 1982 China's relations with the United States turned cooler, in part over the issue of U.S. arms sales to Taiwan. Beijing canceled the provisions for

---

*All dollar amounts are U.S. dollars unless otherwise indicated.

duty-free imports from Taiwan, and relations settled back into a distant mode. A few Taiwanese entrepreneurs, circumventing Taiwan's regulations, made small "indirect" investments in the PRC, beginning in 1983. But these were tiny, experimental projects. The time was simply not ripe for the development of serious economic cooperation.

Taiwan's Economic Reorientation in the East Asian Context

Taiwan's lack of interest in early PRC overtures must been seen in the context of Taiwan's economic position at the time. The low level of Taiwan-PRC economic relations was only one example of the generally low level of inter-Asian trade and investment in the 1970s and early 1980s. A succession of "export miracles" had emerged in East Asia, beginning with Japan, but until the mid-1980s those export and growth successes occurred in the context of relatively low levels of economic integration among the different parts of East Asia. If one examines measures that compare interregional trade to total foreign trade (within and outside the region), they all indicate that East Asian interdependence in trade was lower in the 1970s than it had been in the immediate pre–World War II period.[6] Dependence on outside markets, particularly those in North America, was correspondingly large. For example, Taiwan sent 49 percent of its total exports to the United States in 1984 (the peak year). Thus the near absence of economic relations between the PRC and Taiwan, though it reflected political barriers on both sides, also corresponded to the economic strategies being followed by each side and to its economic interests as it perceived them.

The low levels of interregional trade among East Asian neighbors were an outcome of the economic development strategies that most of those countries pursued. In successive tiers, countries followed policies to promote the export of labor-intensive manufactures. Their own markets—with the exception of Hong Kong—were not very open to imports of manufactured consumer goods.[7] Moreover, the export promotion strategy naturally led to an emphasis on penetrating the U.S. market, since that was and is by far the largest, as well as the most accessible, in the world. Consequently, all the East Asian economies, despite their relative openness as measured by conventional statistics like trade ratios, have in fact been characterized by comparatively modest levels of trade among the (economically similar) neighboring countries and relatively high dependence on the U.S. market. That was certainly true in Taiwan. Export sectors were promoted, but import competing sectors were partly shielded from competition.[8] Although the structure of trade of these exporters might be considered unbalanced, total trade was roughly in balance. None of the newly industrializing economies experienced persistent, large trade surpluses until the 1980s, and Taiwan was no exception. Under these conditions, Taiwan had very modest, if any, interest in fostering economic interchanges with the PRC. Its own economic development strategy was on track.

Taiwan policymaking was extremely forward-looking in recognizing the need to upgrade the quality and technical content of exports. Already by the early 1970s economists and policymakers had begun to recognize the potential vulnerability of Taiwan to competition from up-and-coming lower-wage exporters. At the same time policymakers recognized that whatever policy response was forthcoming had to be consistent with the predominantly small-firm basis of the economy. The approach to the electronics industry was extraordinarily innovative. The government boosted its research and development expenditures and set up the Industrial Technology Research Institute (ITRI) in 1973. Under ITRI the Electronics Research and Services Organization (ERSO) was established the following year, with the mission not only to conduct research but also to diffuse technology to the private sector. For this purpose, ERSO was empowered to set up joint ventures with foreign businesses and private firms to commercialize its technology. Moreover, ERSO was encouraged to transfer technology to small firms on favorable terms. ERSO was notably successful in pursuing this spin-off model, particularly in the integrated circuit (IC) industry, where it was the main source of technology for all Taiwan's major IC producers in the early and middle 1980s. A number of other institutions were set up to encourage the diffusion of technology, including the Institute for Information Industry (III) and the Hsinchu Science-Based Industrial Park, which leased land and provided infrastructure services primarily to electronics firms. Additional financial incentives were provided to electronics firms after the 1980 designation of information technology as a strategic industry.[9]

During the 1980s the gradual accumulation of skills and reputational advantage slowly began to transform the fundamental economic conditions confronted by Taiwan. Human capital accumulation was evident in high and rising education levels, but even more critical was the increased level of knowledge of the international marketplace. Having learned how to export, the Taiwanese applied the knowledge to a broad range of products. Moreover, Taiwan was accumulating reputational advantages as it became clear that Taiwanese manufacturers were reliable, prompt, high-quality suppliers of a broad range of goods. As a result, Taiwan's comparative advantage in a range of labor-intensive manufactures was steadily increasing over time. This fed an explosive process of growth by steadily improving Taiwan's relative cost advantage in exporting.

It is important to stress Taiwan's improving comparative advantage, because it was developing at the same time that economic success was leading to rising living standards and rising wages. Between 1975 and 1985 the nominal wage rate in Taiwanese manufacturing increased at 13.7 percent annually, while nominal labor productivity was growing only half as fast, at 6.8 percent annually.[10] Unit labor costs were growing rapidly. But this did not mean that Taiwan's international competitiveness was eroding through 1985. Rather the opposite: exports grew rapidly and steadily. Maintaining an essentially fixed exchange rate with the United States, Taiwan began to develop trade surpluses after 1970, though these were twice eliminated by oil price shocks in 1974 and 1979. Domestic saving and investment rates both steadily increased through 1980 and were in the range of 30–35 percent of GNP in 1979–81 (figure 3-1). Higher incomes were possible because greater total efficiency in trade (not just in trade-oriented manufacturing) was improving Taiwan's international competitiveness.

During the 1980s the incipient trade surplus Taiwan had been developing suddenly exploded. The trade surplus grew as the U.S. dollar appreciated dramatically following President Reagan's policies of fiscal deficits and tight monetary policy. And surpluses continued to grow in the mid-1980s even as the U.S. dollar lost strength. Oil prices came down, and when the Japanese yen was revalued after the 1985 Plaza Accord, Taiwan's export competitiveness relative to Japan's increased. The current account surplus climbed steadily, reaching the astonishing level of 20 percent of GNP in 1986. This huge disequilibrium is proportionately much larger than any that has ever developed in the Japanese economy. By contrast, the much more gradual adjustment of the Hong Kong economy—made possible in part because of the earlier option to move production into Guangdong—avoided the emergence of dramatic disequilibriums.

Figure 3-1. *Taiwan's Saving, Investment, and the Current Account, 1978–91*

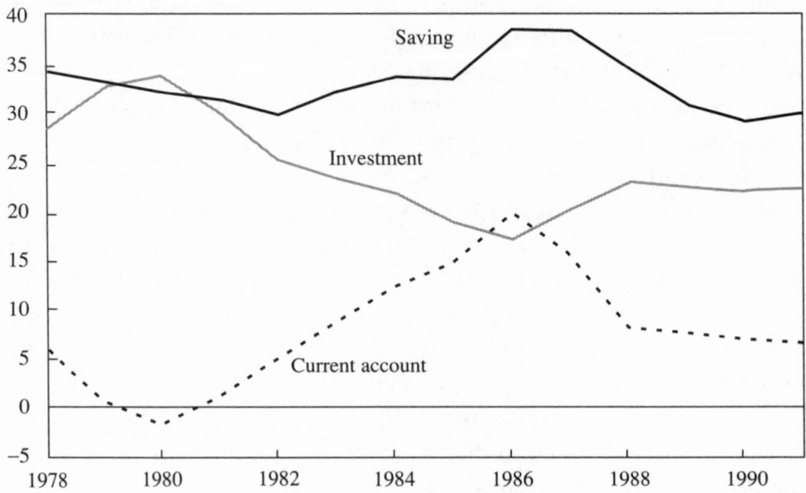

Source: Republic of China, Council for Economic Planning and Development, *Taiwan Statistical Data Book 1990* (Taipei: CEPD, 1996).

The huge Taiwanese surplus naturally put upward pressure on the currency, requiring a revaluation. The current account surplus can also be analyzed by looking at the parallel trends in domestic saving and investment (figure 3-1). Though domestic saving remained high, domestic capital formation declined steadily between 1980 and 1986, falling below 20 percent in 1986. Such trends reflect caution among domestic manufacturers, who may perceive rising costs as ominous portents of declining future competitiveness. More generally, though, declining domestic investment is part of the overall adjustment process. Businesses planning domestic investment may well delay actual investment, because they believe the currency will be revalued upward, reducing the cost of investment goods (which have a high import content). Businesses may treat the surge in export income as transitory, saving large parts of it without making corresponding new investment. Finally, in an economy as open as Taiwan's, businesses will increasingly feel the need to diversify assets into various geographic locations, something that was prevented during 1985–86 by strict foreign exchange controls. Declining domestic Taiwanese investment is part of the total adjustment process, rather than a sign of an exogenous decline in competitiveness that is driving that process.

In any case, the government delayed adjustment by building up foreign exchange reserves (figure 3-2). Rather than adjust the exchange rate, the government ran foreign exchange reserves up from $23 billion to $77 billion between 1985 and 1987. High domestic saving (relative to investment) prevented the government's accumulation of assets from being translated into substantial inflation. Residents of Taiwan accumulated claims against the government while the government accumulated claims against the outside world. Some increase of foreign exchange reserves was probably warranted by Taiwan's peculiar international political position. But as a whole, the increase of reserves meant an excessive diversion of resources into low-yielding assets and thus a significant amount of income forgone. Moreover, by preventing appreciation, the government perpetuated large trade surpluses with the United States that were not politically sustainable.

Figure 3-2. *Taiwan's Foreign Exchange Reserves and Exchange Rate before and after Its 1989 External Adjustment, 1978–92*

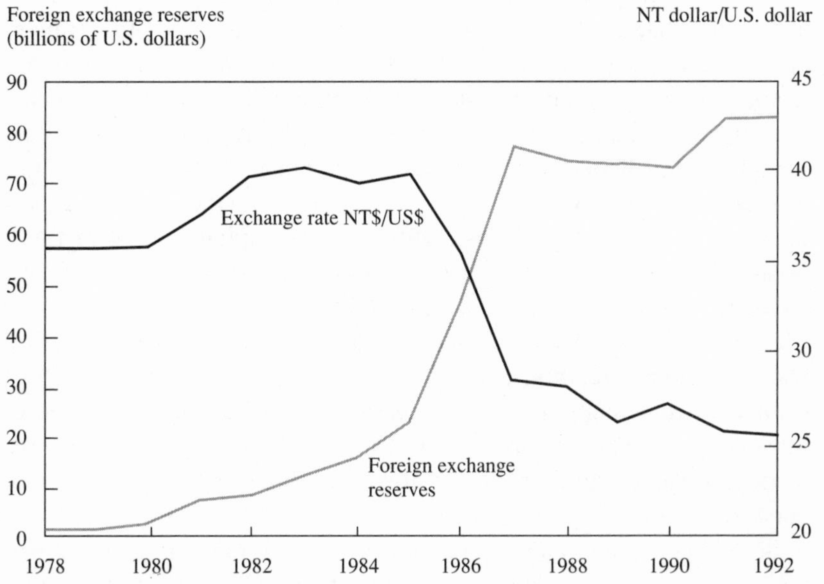

Source: See figure 3-1.

Adjustment finally took place during 1986–87. The currency was revalued upward by 40 percent against the U.S. dollar in two years and has continued to appreciate slightly since. In July 1987, partly in response to U.S. pressure, Taiwan substantially reduced its foreign exchange controls. (But Taiwan avoided large potentially destabilizing capital inflows by maintaining a number of administrative restrictions on domestic capital markets.) Over the next few years adjustment took the form of a combination of a reduced current account surplus—caused largely by currency appreciation—and a steady outflow of private capital. Each of these contributed about half of the overall adjustment required. (Increased government infrastructure spending also contributed to the shrinking of trade surpluses.) Government accumulation of foreign exchange reserves ceased, and the economy reached a new dynamic equilibrium.

This adjustment experience was the basic context within which changing Taiwanese policy toward the mainland needs to be placed. The kind of adjustment Taiwan faced explains both the delay in adjustment and the way in which that adjustment affected Taiwan's relation with the mainland. The big trade surplus and accumulation of reserves created significant political and economic tensions. Yet the reluctance to allow the currency to appreciate can be easily understood. Taiwan had accumulated valuable trade-related capital in a range of light manufactures. Rapid appreciation would destroy Taiwan's trade competitiveness in these products and wipe out much of the accumulated human capital and reputational advantages. Then outside competitors would simply displace Taiwan's existing export networks. Some Taiwanese businesses would successfully make the transition into new export products, but product-specific human capital related to traditional labor-intensive exports would simply be lost. In the face of such concerns, appreciation was reluctant and belated.

However, once it was accepted that appreciation was necessary, Taiwan's economic relation to neighboring regions, including the PRC, changed dramatically. It became imperative for Taiwanese businesses to rapidly move production to low-wage areas, through direct investment. Outward investment would allow Taiwanese businesses to maintain control over export networks and continue to reap a return from trade-related human capital. For the first time Taiwanese businesses needed access to mainland Chinese production locales, or to similar low-cost locales in Southeast Asia. Because of geographic and cultural proximity, the mainland was the cheapest neighboring region. Thus during 1986 Taiwan businesses for the first time developed a strong economic motive for investment in the PRC.

China's Coastal Development Strategy in the Context of Economic Reform

The PRC also undertook a major shift in economic policy in 1986–87. Here the shift can be attributed to disenchantment with the preceding pattern of external economic reform and recognition of the new opportunities that changes in the East Asian economic environment—including rapid adjustment in Taiwan—were bringing. In some respects, then, the PRC's policies can be seen as a response to Taiwan's policy actions and to the broader adjustment process under way in Taiwan.

The PRC had embarked on externally oriented reforms from the beginning of the economic reform process. The overtures to Taiwan described in the first section were a part of this early attempt at an opening. The most prominent sign of early externally oriented reforms was the establishment of special economic zones (SEZs), which date effectively to the early months of 1979 and which were formally established in 1980.[11] The SEZs were important symbolic commitments by the PRC government, and they served as political emblems of the entire reform process. Moreover, they were instrumental in providing foreign businesses with a staging ground from which to manage their entry into the Chinese environment. Finally, the SEZs, and the Shenzhen SEZ in particular, served as a testing ground for reforms, particularly for those relating to price flexibility and labor circulation. For all these reasons the policies of domestic economic reform and external economic opening have frequently been paired as the two constituent elements of the whole reform process.

Although early reforms did not lead to a breakthrough in relations with Taiwan, significant progress on economic reform was clearly a precondition to the development of links with the China Circle. There had to be sufficient autonomy and orientation toward profit in the Chinese domestic economy to make enterprises responsive to the opportunities in externally oriented reforms. Adequate information about market conditions and opportunities had to be available. More specifically, currencies had to be aligned to reasonable values. In the PRC context, this meant that a substantial devaluation had to take place. That was partly accomplished on January 1, 1985, when the posted and "internal" exchange rates were unified at 2.8 yuan to the dollar, representing a real depreciation of about 50 percent from 1980. Currency realignments on both sides of the Taiwan Strait were thus creating a particularly large revaluation of the Taiwanese currency and the Chinese renminbi. Between the beginning of reform in 1978 and 1985, broad-based domestic and external economic reforms had created the general preconditions necessary for further integration in the China Circle.

Reassessment of the Early "Open Door" Policy

Despite these early achievements, by 1985–86 the initial strategy for opening the Chinese economy had largely been exhausted, and Chinese leaders, particularly Premier Zhao Ziyang, the chief economic strategist, were searching for a new and potentially more successful approach. This approach eventually came to be called the coastal development strategy (CDS). This strategy was in important respects a reformulation and repudiation of the initial strategy as envisioned in the SEZs. It was much more radical, and much more successful. The CDS was an alternative not only to remaining fairly closed to the world market but also to having more limited forms of economic opening. As mentioned earlier, China began to open to the world market immediately on the initiation of economic reforms in 1979. But the initial opening was relatively limited. A major increase in exports was envisaged, to ease foreign exchange constraints and increase China's capacity to import technology. However, the policy that was intended to achieve that goal was based on state industry plus a subsidiary role for the SEZs. The initial export promotion strategy was to be a balanced, or "fair," policy, involving efforts by coastal cities, inland cities, and SEZs. Developed coastal cities, especially Shanghai, were to expand exports, building on successes achieved in the 1970s in such sectors as textiles. That would mean ceding much of the domestic market to inland producers. Thus while the advanced coastal areas would have some advantages in terms of access to foreign exchange, they would also have the disadvantage of having to penetrate highly competitive world markets. Inland producers would have little access to foreign exchange, but they would have protected access to the rapidly growing domestic market. Indeed, their growth prospects should be enhanced, because they would face less competitive challenge from the more efficient coastal factories (which were to reorient outward). The role of the SEZ in this program was limited; it was supposed to attract high-tech foreign investment and transfer technology, managerial expertise, and knowledge about foreign markets to domestic producers.[12]

By the mid-1980s this policy package was beginning to unravel. First, export performance had been only moderately successful. Between 1980 and 1985 exports grew at the respectable, but hardly torrid, rate of 8.8 percent annually. Moreover, the composition of exports had changed very little; labor-intensive manufactures grew slightly faster, at 11.4 percent annually, but that was hardly significant. Resource-based products had grown almost as fast as total exports (at 7.9 percent annually), and petroleum, gas, and coal had grown faster than exports as a whole, increasing their share of total exports from 22 to 26 percent.[13] Furthermore, even this result was achieved at a fairly high cost, since export subsidies reached their reform era maximum in 1986.

It was not hard to find the reasons for this lukewarm performance. Instead of leading the export drive, Shanghai and other coastal cities dependent on state enterprises were slowing it down. Manufactured exports produced in Shanghai dropped from 13 percent of national exports in 1978 to only 8 percent in 1985.[14] The SEZs were rapidly increasing exports, as anticipated, but the costs were far exceeding what had been promised. By the mid-1980s the Shenzhen SEZ was importing far more than it exported (five to one by some figures) and was absorbing large government funds for infrastructure creation. Moreover, very little high-tech manufacturing was being attracted into Shenzhen at this time, with only one-third of total projects classified by Chinese analysts as advanced or relatively advanced. Under these circumstances, conservatives opposed to the open-door policy argued that it was time to pull the plug on Shenzhen and leave it to operate without a massive infusion of state resources.[15]

Both Shanghai and Shenzhen were being hampered by the semireformed state of the economy. Shanghai had long been dependent on raw materials produced in other areas of the country. Cotton, wool, tobacco, and other raw materials, procured at low prices, flowed to

Shanghai's factories, where they were processed into high-markup consumer goods. But with the onset of reform, inland provinces were increasingly anxious to keep their raw materials so as to earn lucrative processing fees themselves. Domestic shipments of raw materials to Shanghai stagnated, and in some instances plummeted. Shenzhen's problem was rather different. Incentives to operate in Shenzhen were so great that domestic agents of all kinds poured resources into the SEZ to establish subsidiaries. Economic activity was diverted into Shenzhen from other domestic locations much more rapidly than foreign investment was attracted into Shenzhen. In the short run it appeared that Shenzhen was simply absorbing resources from the Chinese economy.

Indeed, this period of reformulation and reconsideration was important across the board in the economy. As Jean-François Huchet demonstrates in chapter 8, at about this time development strategy for the electronics industry underwent a similar reformulation. With hindsight one can see that Shenzhen's importance was then widely misinterpreted. As a conduit to bring high-tech investment into the Chinese economy, Shenzhen had indeed failed. But its real importance was as a bridge, or window, to a broader economic region, the Pearl River Delta. Shenzhen provided a base of operations and a transfer point for a series of international transactions. The manufacturing in Shenzhen itself was less important than the manufacturing being transferred gradually to the Pearl River Delta as a whole. That is, Shenzhen facilitated both the expansion of Hong Kong manufacturing out of Hong Kong proper and its dispersion throughout the Pearl River Delta. For this process to take place, however, it was essential that the Chinese government lower or eliminate barriers to shipping raw materials or semifinished goods in and out of Hong Kong. The government began to do so as early as 1978, when it allowed various kinds of processing and assembly agreements to be signed between Hong Kong manufacturers and firms (often township and village enterprises) in the Pearl River Delta.

These agreements had important implications. Raw materials and components were transported duty-free from Hong Kong into the delta, processing fees were paid to the local enterprises, and the finished goods were then transported back to Hong Kong. There were already 200 such agreements in 1978, and by 1980 the number had jumped to more than 7,000. These were very small scale projects: the average value of processing fees paid per agreement was only $17,000, so the total value of such fees was only a little more than $100 million in

1980.[16] But the agreements established the principle and the procedures for the duty-free import of materials and components so long as they were used in export production. This was a fundamental measure in developing an export-oriented economy in—and therefore attracting foreign investment to—China's coastal region. But at the time the measure was generally viewed as a minor low-cost adaptation on the fringes of the policy of openness. It took some time for the significance of the measure's success to be apparent; it was not immediately clear that the SEZ policy and the associated abuses were linked to this success. As a result, in the mid-1980s the pro-SEZ policy was still controversial, and conservatives in the Chinese leadership could plausibly argue that the policy had failed.

A Failed Attempt at Trade Liberalization

The Chinese govenment initially responded to the mixed results of the SEZ policy by moving toward a broad liberalization of the trading regime. During 1984 the political pendulum in the PRC swung strongly in the direction of liberalization. Trade reform joined industrial reform at the top of the agenda during this peak period of reformist enthusiasm.[17] Indeed, in some respects, the import regime, as tentatively reformed in 1984, was more liberal than the system that prevailed subsequently in the early 1990s, during the period of rapid expansion of China's trade. However, under the conditions that prevailed in China in 1984–85, that liberalization proved to be unsustainable. Import liberalization led to a rapid increase in imports and a dangerous drawdown of China's foreign exchange reserves. As a result, by mid-1985 Chinese policymakers began backpedaling furiously, reimposing restrictions on the trading process.

The key measures of liberalization during 1984 were the decentralization of import authority to regional trading companies, the attempted movement to the agency system for both importers and exporters, and the opening of fourteen coastal cities to trade and investment.[18] Unfortunately, these measures had very asymmetric effects on imports and exports. With hundreds of new trading companies suddenly gaining the ability to import, imports surged 54 percent in 1985, while exports increased only 5 percent, and the trade deficit surged to $14.9 billion. The regime responded not by canceling the decentralization provisions themselves but rather by imposing a set of new administrative restrictions on importers and exporters. Import licensing, which had already been started in anticipation of the needs of trade decentralization, was extended to a broad range of commodities. A separate program of quantitative restraints on machinery and other capital goods was instituted. Licenses were required for 235 export commodities that had been, in the eyes of Chinese planners, inappropriately exported. Beginning in 1985, access to foreign exchange accounts for exporters was frozen for periods extending as long as two years. The program of opening of fourteen coastal cities was frozen, with opening measures restricted to the four most important cities of Guangzhou, Shanghai, Tianjin, and Dalian. Thus by the end of 1985, the attempted trade liberalization of the preceding year was seen to have failed, and the trading system was hobbled with a new set of administrative restrictions.

This failure did not necessarily mean that the liberalization strategy envisaged was flawed. Indeed, the primary reason that imports surged was the overheating of the domestic economy, which could in turn be traced to problems in implementation of simultaneous reforms in the banking system.[19] But trade problems did reflect the ongoing difficulty of opening up the Chinese economy, given that domestic prices remained far from world prices and that China's foreign exchange earning capacity was still strictly limited. In any case, the short-term failure of the 1984 liberalization meant that Premier Zhao Ziyang had to cast about for a different strategy to move the overall reform process forward.

Changes in the World Market

At the same time changes in the world economy that were shaping other countries' development strategies were also beginning to have an impact on China. Two were most important. The first was the collapse of oil prices in 1986 and the subsequent weakness in all natural resource and commodity prices. China had been a major oil exporter through 1985. Though it had long declared that its policy was to foster manufactured exports, the collapse of commodity prices gave much greater urgency to this quest. The oil price collapse pounded the final nail in the coffin of natural resource–dependent development schemes, in China as in other developing countries. The second external effect was the realignment in currency values that was taking place in Asia. By 1985 the yen's value had already been pushed up by the apparent success of the Plaza Accord. Moreover, during 1986 it began to become clear to Chinese policymakers that Taiwan's currency was also headed for a major appreciation in value. Policymakers in China, like their counterparts in Indonesia and other ASEAN countries, were aware of the opportunities that these currency realignments created (see chapter 4).

Domestic Political Considerations

During the entire period after 1985, Premier Zhao Ziyang was under great pressure to achieve results. The urban reform initiative that had been kicked off in October 1984 had achieved some important results but was well short of a breakthrough. Zhao's initial response, during 1986, was to attempt to develop a coordinated package of reforms that would include reform of prices, taxes, and enterprise finance and incentive structures. In the course of the year, though, it became apparent that it was extremely difficult to achieve consensus on coordinated reforms. Zhao became concerned that his attempt had run into a dead end and become too dangerous to pursue.[20] At the end of 1986 he abandoned the idea of carrying out a coordinated reform of prices and incentives; instead he seized on the coastal development strategy as an alternative, feasible, and potentially successful policy platform. Zhao identified the external sector as an area in which dramatic progress could be made.

In fact, the CDS policy has always been closely identified with former premier Zhao Ziyang. The many controversial elements of the policy (discussed below) have always been debated in a politically charged atmosphere relating to the fortunes of Zhao and his associates. At the end of 1986 the fall of Hu Yaobang, the party chairman, dramatically demonstrated the vulnerability of China's reformers. During 1987 the CDS was a key element of Zhao's domestic political strategy. The CDS could potentially purchase political support from representatives of the coastal provinces, while also delivering a considerable economic benefit to the country as a whole. In addition, the CDS could be seen to combine foreign policy aims with domestic economic aims. Zhao was already using foreign policy (and foreign investment) arguments as a weapon against his domestic conservative opponents in the political maneuvering that followed the fall of Hu Yaobang. If the coastal development strategy could succeed, it would have the additional advantage of fostering unofficial relations with Taiwan, thus giving Zhao a large foreign policy, as well as economic, success.

During 1987 Zhao pushed forward the openness of the Chinese economy to trade and investment. Basically, he extended to the entire coastal area the de facto privileges enjoyed in the Pearl River Delta to import inputs duty-free so long as they were used in export production. Zhao reacted to the problems caused by limited opening and distorted incentives by pushing for further opening. The general principle he advanced was to "use the favorable opportunity created by international adjustments in industrial structure, as well as the favorable conditions created by the abundant labor resources of our coastal areas—among others—to develop in a big way an externally oriented economy in the coastal regions."[21]

The Coastal Development Strategy

The result of these considerations—exhaustion of the existing openness strategy, change in world market conditions, and Zhao's need for a domestic political success—was, as mentioned, the CDS. The policy framework took shape in late 1986 and developed during 1987. Formal announcement and retrospective celebration took place only in early 1988, but such a sequence is common in Chinese politics. The crucial policy elements can clearly be seen to have been put in place in the last few months of 1986 and the first half of 1987. Indeed, the surge in Taiwanese investment that occurred during 1987 was plainly a response to these policy initiatives as well as to the appreciation of the Taiwanese currency.

A few uncommon characteristics of the CDS should be noted at the outset. First, in contrast to many components of Chinese economic reform, the CDS was intended to accomplish certain specific objectives, was implemented with those objectives in mind, and succeeded in achieving those objectives. This result contrasts strongly to the general reform process, which was notable for muddling through and for the persistent importance of unintended consequences. The objective of the CDS was to make the Chinese coast an appropriate site for the reception of labor-intensive manufacturing, in which Hong Kong and Taiwan were losing comparative advantage. This objective was explicitly stated by Zhao Ziyang in early 1988, when the CDS was officially adopted as government policy.[22] The transference to the China mainland of labor-intensive light manufactures for export did not just happen as the result of general economic reforms; instead, a specific package of reforms was adopted for the precise purpose of facilitating that transfer. That package proved to be a remarkable success, perhaps the most outstanding policy design success of the Chinese reform process.

The first stages in the external reform process leading up to the CDS were general reforms in the system of foreign exchange management. These were not only essential to any additional opening of the Chinese economy but also, of course, to the success of the CDS. The first step was the opening of the "swap markets," or foreign exchange adjustment centers, during the course of 1985. These provided a secondary or parallel market on which investors (initially) could trade foreign exchange with minimal official controls over amounts and prices. Subsequently, on July 5, 1986, came another devaluation of the yuan, to 3.7 per U.S. dollar. This brought the real exchange rate down to below 40 percent of its value in 1980, to the point where it reached the approximate real value it retained for the next six or seven years. Finally, in October 1986 the Twenty-Two Articles, or provisional regulations on investment, were issued, which outlined and gave legal sanction to a set of preferential policies for foreign investment that had been informally taking shape over the previous year. These policies, designed to reassure foreign investors who had been troubled by the policy reversals that had characterized the years 1984–85, generally succeeded in this aim. These measures should not be considered part of the CDS per se. Rather, they were part of the broader policy of setting up appropriate economic conditions for further opening that laid the groundwork for the CDS.

The most important specific provisions of the CDS included the following:

—There was a substantial decentralization that gave local governments the authority to establish conditions for foreign investment and trade. In practice, this meant that local governments in coastal provinces obtained the ability to offer tax breaks and concessionary prices for land and other resources.[23] Moreover, localities were given much greater freedom to establish foreign trade corporations (FTCs), which had the right to engage directly in foreign trade. There had been 1,500 FTCs in early 1986, but after 1986 localities were allowed to establish their own FTCs, and the total grew rapidly to 6,000 by 1988.

—Rural enterprises (xiangzhen qiye) were given authority to engage in joint ventures with foreign partners. The attempt to match foreign investors with approved state-sector partners was abandoned. Moreover, the entire coastal area was thrown open to foreign investment.

—Low-tech labor-intensive manufacturing investment was accepted, provided only that the completed projects could earn foreign exchange. The attempt to restrict investment to projects that could promise technology transfer was abandoned.

—The conditions under which foreign-invested firms could import raw materials were further liberalized, thus essentially ending the attempt to base export potential on the development of an integrated domestic market or supply capability. This was the so-called liang tou zai wai (both ends outside) policy—having both supplies and markets outside the domestic economy.

—Preferential policies to foreign investors were consolidated, including, in particular, the de facto legalization of wholly owned subsidiaries of foreign companies. This policy was especially important to Taiwanese investors.

These measures implied the end of the government's attempt to micromanage foreign investment. Furthermore, by greatly expanding the number of potential partners and hosts for foreign investors, the measures substantially reduced the bargaining power of any individual unit on the mainland side. Foreign investors could shop among any number of potential partners to find the best conditions. Competition among participants on the Chinese side would quickly drive conditions toward internationally competitive levels and make further steerage infeasible.

This policy package was enormously successful. Export growth accelerated, and growth of light manufactured exports exploded. The growth rate of manufactured exports doubled, exceeding 22 percent annually for the ten years 1985–95. The share of foreign-invested enterprises (predominantly firms invested by Hong Kong and Taiwan) in total exports expanded rapidly. In 1986 FIEs accounted for just under 2 percent of China's exports; by 1993 they accounted for 26 percent of a total export volume more than three times as large as that of 1986. Even more significant was the growth of exports based on export processing. Export processing involves the duty-free import of raw materials and components, and their processing or assembly into exportables. Such arrangements began as early as 1978 but were small through the early 1980s. By 1988 exports produced by processing duty-free imports had grown to 27 percent of total exports, and the figure expanded further to 45 percent by 1991.[24] Thus the crucial provisions of the CDS are precisely those institutions most closely linked to the acceleration in Chinese export growth after 1985.

Besides benefiting from the main provisions of the CDS, Taiwanese investors were given additional concessionary treatment. They became eligible for tax breaks and qualified for permission to sell 30 percent of output on the domestic market. In 1988 a further package of minor preferential policies was formally adopted by the Chinese State Council. These allowed Taiwan residents to rent or lease government-owned factories, use their mainland relatives as agents, and enjoy certain other privileges. Though none of the measures by themselves reflected an exceptional degree of privilege, together these actions by the central government clearly signaled local governments that they were being actively encouraged to court Taiwanese investors and that they could not get in trouble for cutting a too favorable deal with a Taiwanese counterpart.

The CDS provoked vigorous debate from the beginning. All its measures were controversial, because they implied a substantial relaxation of the government's attempt to control the pace and nature of economic change. Most discussions of the politics of the CDS have focused on the problems of regional imbalance—the fact that policy preferences were given to coastal areas that were already richer, better situated, and (in most cases) more rapidly growing than inland provinces. In addition to those issues, the issues of government control were substantial. Inland provinces objected to the "unfair" advantages they saw Guangdong and other coastal provinces enjoying.[25] Other provisions were also criticized. In 1989 Premier Li Peng took aim at several key tenets. "We do not agree," Li said, "with comrade Zhao Ziyang's inappropriately exaggerating the role of rural enterprises."[26] And in the following year he argued that "Taiwan has people who wish to move 'sunset industries' to the mainland. We have openly expressed the fact that we do not welcome 'sunset industries.' We welcome high technology, and we welcome externally oriented projects that can export."[27] Li failed to recognize that “sunset industries” can be important exporters, appropriate to China’s factor endowments. More significant was his obvious belief that the government can steer foreign investment into specific sectors and vet all investment projects. Thus the CDS was controversial because of its implications for regional issues, ownership issues, and issues of government control. Luckily, the objections of conservatives faded after the astonishing success of the China Circle export economy in the 1990s.

Taiwan's Changing Mainland Policy

The Taiwanese government's policy toward the mainland began to change in 1987, in response both to obvious evidence of PRC reform and to the changed economic situation in which Taiwan found itself. The movement of Taiwanese capital into mainland China beginning in 1987 must therefore be understood in the context both of Taiwan's economic adjustment and of the mainland's reforms, especially the coastal development strategy. The long-delayed currency appreciation played the catalytic role. As Taiwan's currency appreciated, the economic interests of Taiwanese businesses shifted sharply. It began to be strongly in the economic interest of Taiwan to permit businesses to relocate to the mainland. At this point—and only at this point—did the Taiwanese government begin to adopt policies to allow Taiwanese businesses to operate on the mainland.

During 1987 Taiwan adopted a number of complementary policies to permit investment and trade in the mainland. These were not the first policies: in 1985 the Taiwan authorities had adopted an explicit noninterference principle toward indirect exports to the PRC (that is, exports through Hong Kong), and this principle was restated during 1987. Two new measures were particularly important. First, the Taiwanese government lifted martial law and permitted Taiwanese citizens to visit the mainland. Second, the need for Central Bank approval for outward remittances of capital was removed for amounts below $5 million. Thus, for the first time, indirect movement of goods, people, and money through Hong Kong to the PRC was permitted.

Taiwan still hoped to use its economic weight to achieve leverage with the PRC. An important way to do so so was to maintain a "negative list" approach to regulating business with the PRC. Product lists were established: if an item was not listed as being approved for business in the PRC, it was to be considered forbidden. Investment in the PRC was gradually permitted for more than 3,000 specific products. Progress in liberalizing permitted imports into Taiwan was slower. In

1987 the Taiwanese government allowed thirty categories of PRC raw materials (but no manufactured goods) to be imported into Taiwan. This list has only gradually been broadened, mostly during the 1990s. By the summer of 1994 slightly more than 5,000 items—out of a total customs list of 9,000 items—have been approved for foreign trade with the mainland.

What were the consequences of these policies for Taiwan? The question should be answered with respect to both economic restructuring and Taiwan's political objectives. First, the Taiwanese economy could not have adjusted as successfully as it did if it had not had the option of moving production onto the mainland. Thus Taiwan's investment in the PRC is an integral part of its progression to a higher stage of economic development and of its adjustment to the disequilibriums caused by delays in making that adjustment.

Second, Taiwan has built on its success in adjusting to the initial trade disequilibriums of the mid-1980s by moving to a further success in technological upgrading. It has been able to push technological capacities upstream to develop significant research and development and innovation capability. Research and development expenditures in Taiwan increased by a multiple of five between 1980 and 1990, rising from $500 million to $2.5 billion. Government played a key role in this process. Government funding of the Industrial Technology Research Institute has been used as seed money to encourage increased private funding of R&D. Consortia of private firms, which may include some joint ventures with government agencies, and which enjoy tacit government support, have played an important role in upgrading manufacturing technology. Overall, the ratio of R&D expenditure to GDP has increased steadily from 0.8 percent in 1979, when data were first calculated, to 1.8 percent in 1994, the most recent data available. In recent years, R&D funding has come about half from government and half from private industry.[28] The resources put into R&D in Taiwan have been reflected in Taiwan's solid position in a number of niches within the computer industry.

The technological success that lies behind this achievement is amply documented in statistics on U.S. patents granted. During the 1983–90 period residents of Taiwan experienced the highest growth rate in patents granted, at 41 percent annually. Moreover, the types of patents granted to residents of Taiwan shifted strongly toward the more technical classes, including communications technology, semiconductor manufacturing, and internal combustion engines. By 1990 residents of Taiwan accounted for slightly less than 1 percent of all U.S. patents granted. Although well below the 22 percent of total patents granted to residents of Japan, residents of Taiwan received three times the number granted to residents of Korea, and fourteen times the number granted to Hong Kong residents.[29]

Taiwan's Quest for International Recognition

Although relations with the mainland have led to an unqualified economic success, Taiwan's attempts to leverage those relations to produce political benefits have generally failed. As economic factors increasingly gave Taiwanese businesses an interest in cooperating with the mainland, the Taiwanese government's ability to use economics as a bargaining chip for political recognition faded. Repeatedly, Taiwan has tried to delay some particular step in improved coordination with the mainland in order to obtain mainland concessions on the core issue of political sovereignty. Repeatedly, Taiwan has failed. In May 1988 Taiwan returned to the Asian Development Bank, after several years' absence in protest of the bank's decision to change Taiwan's name from the official Republic of China to the "Taipei, China," formulation preferred by Beijing. In some ways a more fundamental concession occurred on March 9, 1991, when Taiwan established an "unofficial" body to coordinate dealings with the mainland. This body, the Straits Exchange Foundation, enabled important talks between the mainland and Taiwan to take place under the aegis of the Straits Exchange Foundation and Association for Relations across the Taiwan Straits, the counterpart "informal" organization set up by the PRC. In all these cases, flexibility and a willingness to compromise were displayed by both sides. But the fundamental concession was made by Taiwan, which agreed to allow discussion of substantive issues with the PRC in a forum in which Taiwan's sovereignty was not officially recognized. At the same time the PRC made no significant concession on the core issue of sovereignty recognition. Indeed, the PRC has never made any move that has explicitly recognized Taiwan in any way as a national entity possessing the attributes of sovereignty.

Taiwan's inability to translate its economic assets into political recognition may appear surprising. After all, Taiwan possesses most of the valuable economic assets—including capital, technology, experience, and reputation—whereas China offers primarily its low-cost labor and access to its market. But since in the relationship between the two sides, private firms in Taiwan are able to capture most of the benefits of economic cooperation, it is difficult for the Taiwanese government to hold up the expanding economic exchanges to pressure the mainland authorities. As a result of the unequal distribution of bargaining power, Taiwan was being pushed into increasingly close quasi-official relations with China, without extracting any recognition of its sovereignty demands.

In the early 1990s, however, Taiwan's international position was being altered by three factors. First, Taiwan had emerged as a major source of investment in the East Asian region. As Tan shows in chapter 4, Taiwan was one of the most important investors in several ASEAN countries as well as in mainland China. This investment outflow naturally gave Taiwan additional diplomatic influence among investment host countries. Second, Taiwan made a transition to a democratic government, a transition marked by legislative elections in 1995 and culminating in the direct election of the president in 1996. Taiwan's successful transition to democracy naturally increased its international influence, because it became harder for countries to ignore Taiwan's claims to recognition. The third factor strongly reinforced the second: beginning in 1991 Taiwan progressively abandoned its claim to sovereignty over the mainland, which it had maintained since 1949. As Taiwan dropped its claim that it possessed legitimate authority over the mainland, mutually exclusive with the PRC government's legitimate authority, it developed a more plausible and persuasive claim to exercise legitimate authority on the island of Taiwan. Taiwan began to argue that there were two separate legitimate governments, which could coexist and deal with each other as equals. Coming from a democratically elected government, such a claim was hard for third-party countries to ignore.

Thus by the mid-1990s Taiwan found itself in a peculiar position. In its relations with the mainland, despite its lucrative economic relations, it was trapped in a situation in which it was unable to leverage its economic assets and improve its political bargaining position. At the same time its potential international assets were growing. Taiwan responded to this situation with a dual initiative. First was an energetic and explicit campaign to "increase international space." It launched a program of informal "vacation diplomacy," which was, unsurprisingly, targeted initially at the ASEAN countries that were emerging as important recipients of Taiwan's investment (and competitors of the PRC). A satisfyingly respectful and businesslike reception was in fact enjoyed by the Taiwanese president in Southeast Asia. Thus Taiwan was attempting to go outside the bilateral relationship with the mainland to develop some additional bargaining power in pursuit of its core objectives of legitimacy and international recognition. Second, in mid-1994 the Mainland Affairs Commission working group resolved not to separate politics and economics in the future and to suspend further relaxation pending positive indications from the PRC.

These maneuvers were profoundly threatening to the PRC. The Beijing leaders were, after all, satisfied with the status quo. They had managed to achieve intimate economic ties and rapid economic growth without any serious concessions. But if Taiwan was able to achieve a sudden breakthrough in its quest for international space, this would, from the perspective of Beijing, upset the entire apple cart. Under the status quo the basic framework of relations incorporated the asymmetry between the two sides. The one-China policy implied that some day China and Taiwan would be reunited under the aegis of a future Beijing government. The Beijing government therefore enjoyed the sole fully legitimate position in the political order. But if Taiwan could really achieve dual recognition for Beijing and Taipei, that structure would be shattered. When Li Teng-hui, the president of Taiwan, was allowed to "unofficially" visit the United States in 1995, the prospect of a sudden overturning of the framework for political relations all at once seemed possible, perhaps even likely, to the Beijing leaders. Their discomfort was intensified because they were not notified of the visit until the day it was approved, thereby creating suspicion of collusion between the United States and Taiwan.

In response to these events the PRC began the lobbing of armed missiles into the ocean around Taiwan, while engaging in bellicose rhetoric and a level of polemic not seen in many years. These actions by the PRC had enormous negative consequences on its reputation and image in the Pacific region. But such actions should be seen primarily as the most recent move in a complex set of maneuvers between the two rivals across the Taiwan Strait. The response of the PRC was extreme because it was attempting to draw a line in the sand, limiting behavior within a policy framework that had already developed. By reacting in an extreme and irrational manner, the PRC leaders signaled clearly that they could be expected to behave as partners only within the framework already established.

As of the middle of 1997 the Beijing government seemed to have succeeded in this effort. Taiwan's attempt to break out of the framework of relations maintained by Beijing has, at least temporarily, failed to persuade any significant third party to break with the status quo. Instead, the Taiwanese president has had to resort to the former policy of threatening to withhold government approval for large-scale investments on the mainland on the ground that they threaten Taiwan's national security. Such a threat implies a tacit acceptance of the weak bargaining position in which Taiwan was placed before the eruption in relations. Moreover, in December 1996 the PRC was able to convince the new government of South Africa to shift recognition to Beijing, thus spiriting away one of the most important countries that had still maintained official diplomatic relations with Taiwan. For the present at least, Beijing has managed to preserve the status quo, though at considerable cost to itself.

Conclusion

Both Taiwan and the PRC have benefited enormously from the growth of their economic interactions. On both sides the governments have taken the necessary steps to adjust their economic policies and follow flexible and realistic diplomatic strategies. The PRC acted with uncharacteristic boldness and vision in adopting the coastal development strategy, and it continued to reap the economic rewards from that strategy long after its architect, Zhao Ziyang, had fallen from power. On the Taiwanese side response to the need to restructure the economy was sluggish, and response to the mainland's economic initiatives was dilatory and mildly obstructionist. Yet despite these shortcomings, in the final analysis the Taiwanese government was able to adapt policy quickly enough and profoundly enough to permit the Taiwanese economy to undergo a vital round of restructuring and upgrading. Moreover, Taiwan's subsequent and related domestic economic initiatives have built firmly on that foundation.

At the same time the success of the economic relationship between the PRC and Taiwan has intensified the disproportion between Taiwan's economic strength and its minimal political and diplomatic presence. The economic relation with the PRC has been one of the means through which Taiwan has achieved a substantially higher standard of living and qualitatively superior technological performance. Simultaneously, social development has been accompanied by democratization and the creation of a political and economically more open society. The imbalance between economic and political achievement, on the one hand, and external recognition, on the other, creates a potential for instability in the future evolution of the China Circle. Essentially, it would take a shift in the bargaining objectives of the PRC to resolve the imbalance in economic and political relations.

Notes

1. Because I focus in this chapter on economic policy, I do not attempt to bring out the full complexity of relations between Taiwan and the PRC. Much of that complexity is described with great subtlety in Tun-jen Cheng, Chi Huang, and Samuel S. G. Wu, *Inherited Rivalry: Conflict across the Taiwan Straits* (Boulder, Colo.: Lynne Rienner, 1995).

2. Cyril Lin, "Open Ended Economic Reform in China," in Victor Nee and David Stark, eds., *Remaking the Economic Institutions of Socialism: China and Eastern Europe* (Stanford University Press, 1989), pp. 95–136; Wang Xiaoqiang, " 'Groping for Stones to Cross the River': Chinese Price Reform against 'Big Bang,' " University of Cambridge, Department of Applied Economics, 1993; and Barry Naughton, *Growing out of the Plan: Chinese Economic Reform, 1978–1993* (Cambridge University Press, 1995), pp. 22–24.

3. Stephan Haggard, *Pathways from the Periphery* (Cornell University Press, 1989). See also Robert Wade, *Governing the Market: Economic Theory and the Role of Government in East Asian Industrialization* (Princeton University Press), 1990.

4. Adherence to this position by the Taiwan government had the inadvertent effect of making it easy for countries to switch their formal diplomatic recognition to the People's Republic of China. Governments forced to chooose between the two found it fairly easy to decide that the PRC looked more like a government of all China than did the Republic of China on Taiwan. The policy was set when France extended formal diplomatic recognition to the PRC in 1964, and Taiwan automatically broke relations with France, even though it was unclear at that time whether the PRC would have insisted on de-recognition. François Joyaux, "Le nouveau triangle Paris-Pekin-Taipei," *Politique Internationale* 61 (Autumn 1993), pp. 1–14. One of the many benefits of democratization in Taiwan has been that it enabled the Taiwan government to formulate a much stronger defense of its right to recognition as a sovereign entity. See David S. Chou, "ROC's Struggle for Recognition," *Free China Journal*, July 26, 1996, p. 7.

5. Ralph Clough, *Reaching across the Taiwan Strait* (Boulder, Colo.: Westview, 1993), p. 41.

6. Peter Petri, "The East Asian Trading Bloc: An Analytical History." in Jeff Frankel and Miles Kahler, eds., *Regionalism and Rivalry: Japan and the United States in Pacific Asia* (University of Chicago Press, 1993), pp. 21–48.

7. During the early 1980s Taiwan's imports consisted of 77 percent agricultural and industrial raw materials, 15 percent capital equipment, and only 8 percent consumer goods. By the early 1990s the share of consumer goods had increased to 12 percent, and the share of capital equipment to 17 percent, so that agricultural and industrial materials had declined to 71 percent. *Chunghua minkuo chingchi nienchien,* 1993 (Economic yearbook of the Republic of China) (Taipei: Chingchi Jipao), p. 880.

8. Tyler S. Biggs and Brian Levy, "Strategic Interventions and the Political Economy of Industrial Policy in Developing Countries," in Dwight Perkins and Michael Roemer, eds., *Reforming Economic Systems in Developing Countries* (Cambridge, Mass.: Harvard Institute for International Development, 1991), pp. 373, 377.

9. Chi-ming Hou and San Gee, "National Systems Supporting Technical Advance in Industry: The Case of Taiwan," in Richard R. Nelson, ed., *National Innovation Systems: A Comparative Analysis* (Oxford University Press, 1993), pp. 396–413; Nagy Hanna, Sandor Boyson, and Shakuntala Gunaratne, "The East Asian Miracle and Information Technology: Strategic Management of Technological Learning," Discussion Paper 326 (Washington: World Bank, 1996), pp. 121–38; and Kenneth Kraemer and others, "Flexibility, and Policy Coordination: Taiwan's Computer Industry," *Information Society,* vol. 12 (Spring 1996), pp. 215–49. In the first half of 1995, 96 percent of the output of firms in the Hsinchu Science Industry Park was from producers of ICs, computer products, and telecommunications equipment. Science Industry Park Management Office, "Survey of Firms in the Science Industry Park," in *Diaocha tongji zixun jibao* (Quarterly report of surveys and statistics) (Taipei: Xingzhengyuan Zhujichu, 1996), pp. 26–30.

10. Jung-feng Chang, *T'ai-hai liang-an ching-mao kuan-hsi* (Economic and trade relations across the Taiwan Straits) (Taipei: Kuo-chia Cheng-ts'e Yan-chiu Tsz-liao Chung-hsin [Institute for National Policy Research], 1989), p. 34.

11. George T. Crane, *The Political Economy of China's Special Economic Zones* (Armonk, N.Y.: M. E. Sharpe, 1990).

12. Yu-Shan Wu, "Economic Reform, Cross-Straits Relations, and the Politics of Issue Linkage," in Cheng, Huang, and Wu, *Inherited Rivalry,* pp. 117–18.

13. World Bank, *China: Foreign Trade Reform* (Washington, 1994), p. 9.

14. Shanghai Statistics Bureau, *Shanghai tongji nianjian* (Shanghai statistical yearbook), 1991, p. 326; and Nicholas Lardy, *Foreign Trade and Economic Reform in China, 1978–1990* (Cambridge University Press, 1992), pp. 126–27.

15. Fuhwen Tzeng, "The Political Economy of China's Coastal Development Strategy: A Preliminary Analysis," *Asian Survey,* vol. 31 (March 1991), p. 274.

16. Joseph Chai, "Industrial Co-operation between China and Hong Kong," in A. J. Youngson, ed., *China and Hong Kong: The Economic Nexus* (Oxford University Press, 1983), pp. 108–12.

17. Naughton, *Growing out of the Plan,* pp. 173–80.

18. Information on this episode can be extracted from the general accounts of trade reform in Nicholas Lardy, *Foreign Trade and Economic Reform in China, 1978–1990* (Cambridge University Press, 1992), pp. 39–46; Jude Howell, *China Opens Its Doors: The Politics of Economic Transition* (Boulder, Colo.: Lynne Rienner, 1993), pp. 44–80;

and World Bank, *China: Foreign Trade Reform* (Washington, 1994), p. 3 and passim. The specific account of the failed 1984 liberalization draws heavily on Jiang Xiaojuan, *Jianruo "fuguan" chongji de guoiji jingyan bijiao* (A comparison of international experience in reducing the adverse impact of entering GATT) (Beijing: Jingji Guanli, 1995), pp. 28–32.

19. Naughton, *Growing out of the Plan,* pp. 253–59.

20. Naughton, *Growing out of the Plan,* pp. 173–99.

21. "Zhao on New Economic Plan for Coastal Areas," Xinhua News Service, January 22, 1988; translated in Federal Broadcast Information Service (FBIS), January 25, 1988, pp. 10–15.

22. "Zhao on New Economic Plan."

23. Richard Pomfret, "Taiwan's Involvement in Jiangsu Province: Some Evidence from Joint-Venture Case Studies," in Sumner La Croix, Michael Plummer, and Keun Lee, eds., *Emerging Patterns of East Asian Investment in China: From Korea, Taiwan, and Hong Kong* (Armonk, N.Y.: M. E. Sharpe, 1995), pp. 167–78.

24. Barry Naughton, "China's Emergence and Prospects as a Trading Nation," *Brookings Papers on Economic Activity 1: 1996,* pp. 298–302.

25. Dali Yang, "China Adjusts to the World Economy: The Political Economy of China's Coastal Development Strategy," *Pacific Affairs,* vol. 64 (Spring 1991), pp. 42–64; and Tzeng, "Political Economy of China's Coastal Development Strategy," pp. 270–84.

26. Li Peng, "Correctly Understand the Current Economic Situation and Do an Even Better Job of Rectification," *Qiushi,* no. 21 (1989), pp. 4–5.

27. Li Peng, "Make an Effort to Get the Market Moving: Create an Appropriate Development of Production," *Renmin Ribao* (Haiwai ban), October 9, 1990.

28. Bob Johnstone, "Taiwan: Past Success Provides No Sure Guide to the Future," *Science,* no. 262 (October 15, 1993), pp. 358–60; Hou and San, "National Systems Supporting Technical Advance," p. 407; and Council for Economic Planning and Development, *Taiwan Statistical Data Book,* 1996 (Taipei, 1996), p. 103.

29. National Science Board, *Science and Engineering Indicators,* vol. 10 (Washington, 1991), pp. 133–64.

CHAPTER FOUR

China and ASEAN: Competitive Industrialization through Foreign Direct Investment

Tan Kong Yam

UNTIL THE EARLY 1980s China and many of the Association of Southeast Asian (ASEAN) countries clung to import-substituting industrialization strategies and remained restrictive in trade and foreign investment policies. Protection of domestic industries under this strategy made possible the continuation of state-sponsored industrialization (in China) and the proliferation of inefficient public enterprises (in ASEAN countries). The role of market forces and the potential role of foreign direct investment (FDI) in fostering dynamic economic growth and development were circumscribed in both cases. During the 1980s both regions made a transition to much more open trade and investment regimes, accelerating economic growth significantly.

By the mid-1980s China and most ASEAN countries had made the fundamental changes needed to sustain more open trading regimes. This parallel experience can be partly attributed to the fact that policymakers in both regions were responding to a similar set of external conditions. Policymakers subscribed to similar interpretations of the success of the four East Asian newly industrializing economies (NIEs), comprising South Korea, Taiwan, Hong Kong, and Singapore, and gradually came to hold similarly positive assessments of the impact of FDI. Changes in relative prices—including prices of mineral resources and real exchange rates—pushed policy changes during the crucial years of the mid-1980s. New opportunities were created in all these countries by the increasing need for restructuring faced by industries in the more advanced Asian economies.

At the same time parallel policy evolution created intense rivalry and competition between China and the major ASEAN countries, a rivalry that persists to the present day. Given comparable endowments and levels of development, similar policy regimes place China and most ASEAN countries into direct competition for markets, especially for incoming foreign investment. In this rivalry, advantage has seesawed between China and the ASEAN group. This competition is likely to continue, and perhaps intensify, in coming years.

East Asian Growth and Foreign Direct Investment

After World War II ended in 1945, the General Agreement on Tariffs and Trade (GATT) was established to create an open and free trading system. Together with the system of fixed currency exchange established under the International Monetary Fund (IMF) agreement at Bretton Woods, a stable global trading and monetary system was instituted. These two key postwar institutions, sustained by the hegemonic power of the United States, ushered in an unprecedented period of steady expansion in world output and trade and closer economic interdependence in the noncommunist free world. World trade (volume terms) expanded at an average annual rate of 5.6 percent between 1953 and 1963 and 8.5 percent between 1963 and 1973, much higher than the average rate of 3.5 percent between 1873 and 1913 and the 0.9 percent in the interwar period of 1919 to 1939.

This unprecedented expansion in world output and trade in the postwar era provided Asia-Pacific economies such as Japan and the four East Asian NIEs with a conducive and stable environment for export-led growth. They were lucky to set sail on the stream of industrial catch-up when the gust of wind was strongest. Consequently, during the past three decades the four NIEs have become the most dynamic middle-income economies in the world. Their annual growth rates in GNP per capita between 1965–90 averaged 6 to 8 percent, almost triple the average rate of 2.3 percent for middle-income economies of the world and double the 3.6 percent average for countries in ASEAN, excluding Singapore.

The pattern of industrialization and exports was similar in all the NIEs. After a short period of protectionist import substitution policies in the 1950s and early 1960s, they soon turned to an export-oriented strategy for growth. During the early stage of the outward-oriented development strategy in the 1960s, the emphasis was on the production and export of traditional labor-intensive products such as textiles, clothing, footwear, toys, leather goods, and other light manufactured goods. Technology for these products was standard, the main competitive factor in the world market being low labor costs relative to productivity. The NIEs' major markets were the OECD countries, particularly the United States, which had a great interest in nurturing these market economies, through generous foreign aid, capital inflow, technology transfer, market access, and special tariff preferences, against the ideological challenge of the socialist countries of East Asia (China, North Korea, and North Vietnam). Consequently, manufacturing exports of the four NIEs were able to grow from 20 to 50 percent a year during 1965–73, and despite the first and second oil shocks, they grew at an annual rate of 13 to 21 percent between 1973 and 1985. Rising income and savings, combined with higher educational levels and better infrastructure facilities, allowed the NIEs to invest in and upgrade to more capital-, skill-, and technology-intensive industries such as steel, shipbuilding, electrical machinery, telecommunications, and office automation equipment.

Benefiting from being latecomers to the process of industrialization and taking full advantage of the world trading system, the NIEs managed to complete in 20 to 30 years an industrialization process that had taken the OECD countries 100 to 150 years to accomplish in the nineteenth and early twentieth century. Unlike Japan, however, the NIEs, and particularly Singapore, industrialized with considerable dependence on FDI. Throughout the 1980s the exports of foreign firms accounted for about 80 percent of total exports for Singapore, 25 percent for Korea, 15 percent for Taiwan, and 20 percent for Hong Kong.

By the beginning of the 1980s the success of the NIEs had begun to influence the views of policymakers in ASEAN countries and China. They increasingly looked on the inflow of FDI within a liberal trading regime as a quick way to jump-start the process of industrialization. The key advantages of FDI are that it brings in knowledge of new product and processes, managerial and marketing expertise, and entrepreneurial skills. Developing countries liberalize their foreign investment regimes and seek FDI not just to obtain capital funds and foreign exchange to make up for resource gaps. More important, attracting FDI is a dynamic and efficient way to secure much-needed industrial technology, managerial expertise, marketing know-how, and networks to stimulate domestic industries, as well as to improve national growth, employment, productivity, and export performance.[1]

In 1978, after eighteen years of collaborating with foreign multinational corporations (MNCs) in industrialization, the prime minister of Singapore summarized a study of Singaporean export-oriented industrial firms set up since 1960: "When Singaporeans went into joint ventures with US, European or Japanese foreign entrepreneurs who provided the know-how, the experience, and the marketing, their casualty [failure] rate went down from 38 percent to 7 percent, just 1 percent higher than the 6 percent failure rate of the wholly foreign enterprises. When Singaporeans had less advanced partners from Hong Kong and Taiwan, their failure rate was 17 percent. . . . Learning from scratch in the Singapore experience proved a costly business."[2]

John Dunning's eclectic theory of FDI identified three requirements for direct investment to be superior to alternative modes of business arrangements.[3] First, the investing firm must have an ownership advantage over competitors in the host country. The ownership advantage can arise from a product or process monopoly, a unique or superior technology, better knowledge of the market, or better marketing technique. Second, the host country must possess locational advantages to attract investments. That could be a large actual or potential domestic market (particularly if market access is restricted by existing or impending protective barriers), a strategic regional location, or a cost-effective export production base with abundant low-cost labor, low transportation costs, weak exchange rates, special tariff preferences, generous investment incentives, and possibly lax pollution, labor, or environmental standards. Third, there must also be an internal advantage that induces the investing firm to choose direct investment over other arrangements (production licensing or franchising).

In examining the determinants of developing-country growth in seventy-eight countries during 1960–85, Blomstrom, Lipsey, and Zejan found that the inflow of FDI had a significant positive effect on income growth rates.[4] However, the positive effect seemed to be confined to the higher-income developing countries and was not evident among the poorer developing countries. These results imply that inward FDI becomes a source of more rapid growth only when the host country attains sufficient development and skill to allow the country to absorb new technology and become imitators of or suppliers to the MNCs. Otherwise, FDI is likely to result in enclaves with limited spillover effect to the industrial development of the host country. Preliminary evidence indicates that the growth effect of FDI is significant in China as well.[5]

With increasingly strong evidence of FDI's effect in powering the growth of industrial output and its attendant spillover effect in technological, managerial, and marketing know-how across firms in the host country, it is not surprising that ASEAN countries like Indonesia, Malaysia, Thailand, and the Philippines (as well as China) are actively pursuing FDI to hasten their industrial catch-up process. With a growing global scramble for scarce capital, host countries increasingly must compete aggressively for FDI by providing physical infrastructure, institutional support, pro-business bureaucratic procedures, human resources and labor, and generous tax incentives, as well as by minimizing restrictions and other onerous performance requirements.

Investments in Southeast Asia, 1987–90

The substantial decline in oil and commodity prices between 1982 and 1986 resulted in a significant deterioration in the terms of trade for ASEAN countries. Prices (in current U.S. dollars) of nonfuel primary products fell 7 percent in 1981 and another 12 percent in 1982. More specifically, rubber prices fell by 23 percent between 1981 and 1985, while tin and palm oil prices fell by half between 1981 and 1986. As a result, the Malaysian terms-of-trade index fell by about 24 percent between 1980 and 1986. In addition, petroleum prices fell steadily from about $39 a barrel in 1981 to a low of $14.8 a barrel in 1986.[*] These external shocks to the economy were severe for commodity-based ASEAN economies. The World Bank estimated that because of external disturbances during 1983–88, Indonesia suffered an income loss equivalent to 9 percent of its annual GDP. The financial and budgetary burdens of inefficient state-owned enterprises, masked during the commodity boom in the 1970s, were starkly revealed. This dra-

*All dollar amounts are U.S. dollars unless otherwise indicated.

matic decline in the terms of trade and consequent pressure on the budget energized the political will for liberalization and assault against vested interests and inefficiencies.

These economic and financial liberalization policies in ASEAN were aimed at directing economies toward a system of regulation based on the competitive markets of the private sector. The basic paradigm is the preeminence of competition, whether domestic or international. The key areas of liberalization are international trade in goods and services, including tariff, quota, and licensing structures; internal and external capital movement, including the abandonment of financial repression, deregulation of interest rates, and easing of restrictions on foreign banks and other financial institutions; and competitive goods and factors markets, including the suppression of domestic rent seekers, elimination of subsidies, and significant revision and liberalization of the foreign investment regulations (such as equity rules on foreign investment, business fields open to foreign investors, and local content regulations).

The Special Action Committee on Incentives in Malaysia helped to clarify and harmonize tariff protection. Customs duties for agricultural products, intermediate goods, and capital goods were lowered, and licensing requirements were greatly liberalized and bureaucratic delays reduced. In Indonesia, protectionist barriers were diminished substantially, particularly for imports used by export-oriented producers. An across-the-board reduction in nominal tariffs was implemented in March 1985. In May 1986 measures to provide internationally priced imports to exporters were announced. Through a series of measures in October 1986, January and December 1987, and November 1988, import licensing restrictions were reduced. Steps taken in December 1987 lessened the anti-export bias of trade policy by reducing regulatory restrictions for exporters. Nontariff barriers were replaced with tariffs, while the average tariff rate was lowered. As a result, the value of imports subject to controls declined from 43 percent in mid-1986 to 21 percent in December 1988, and the proportion of domestic production protected by import licensing restrictions declined from 41 to 29 percent in the same period. This reform process continued to May 1990, when import licensing restrictions on 335 products (extending across cement, fertilizers, synthetic yarn, machinery, and electronic products) were removed, and tariff rates were lowered from an average of above 60 percent to below 40 percent.

In Malaysia, regulations governing foreign direct investment also were eased. Under the industrial master plan (1986–90), which called for an outward-oriented industrialization strategy, the private sector was to have an increasing role in development. The administratively cumbersome Investment Incentives Act was replaced in 1986 by the Promotion of Investment Act, which also provided additional tax incentives. Foreign investment was greatly encouraged through the easing of ownership rules and the establishment in October 1988 of a Coordination Center on Investment at the Malaysian Industrial Development Authority as a one-stop agency for foreign investors. In 1988 foreign holdings could attain 100 percent of the equity capital of an enterprise that exported at least 80 percent of its output or was part of the tourism sector.

In Indonesia, the government greatly simplified the list of industries from which foreign investors were barred. Through a series of steps between May 1986 and December 1987, it streamlined the investment approval process, reduced licensing requirements, curtailed the bias against foreign investment, and diminished the role of local content. In May 1989 the government converted the investment priority list to a short negative list, greatly increasing the transparency of the system. In addition, reforms in the area of customs, ports, and maritime transport implemented since 1985 resulted in a significant cut in procedural time and a substantial reduction in business costs. Equity rules of foreign investors in export-oriented manufacturing were relaxed, and foreigners were allowed to buy up to 20 percent of shares in domestic companies. Local content regulations were also relaxed. Foreign firms setting up in Batam were allowed 100 percent equity in export-oriented industries, resulting in an explosion in investment interest from Singapore-based MNCs and Singaporean firms.

These economic liberalization measures in ASEAN were somewhat similar to those carried out extensively in South Korea and Taiwan between 1960 and 1965, when the two countries shifted from import substitution to export promotion. As the return to private sector mechanisms began to work, such deregulatory policies further convinced ASEAN governments to relax and reform policies that bring in more export-oriented foreign investments and stimulate the domestic private sector.

More significant, at the same time as these ASEAN countries began to learn and absorb the lessons of the NIEs' success and were keen to liberalize and deregulate their economies to welcome the inflow of foreign capital and technology, there was an outflow of foreign direct investment from the NIEs and Japan. These Northeast Asian countries, as the result of U.S. and European Community protectionism, rising domestic wages, and the strengthening of currencies (figure 4-1), were searching for cheaper offshore production bases to sustain their international export competitiveness and profitability. They found them in the ASEAN countries. In particular, since 1986 the significant appreciation of the Japanese yen, Taiwan dollar, and Singapore dollar against the depreciating Indonesia rupiah and Chinese yuan has been the main factor in attracting FDI from Japan and the NIEs into ASEAN and China.

Consequently, after the mid-1980s, flows of FDI from Japan and the NIEs to ASEAN were very rapid. Between 1986 and 1990 approved foreign direct investment in Malaysia rose twelve times, to $6.2 billion; twenty-four times in Thailand, to $14.1 billion; and eleven times in Indonesia, to $8.8 billion (table 4-1).[6] By 1990 the major foreign investors in ASEAN countries were Taiwan, Japan, Korea, Hong Kong,

Figure 4-1. *Indexes of Selected Asian Currencies against the U.S. Dollar, 1985–95*

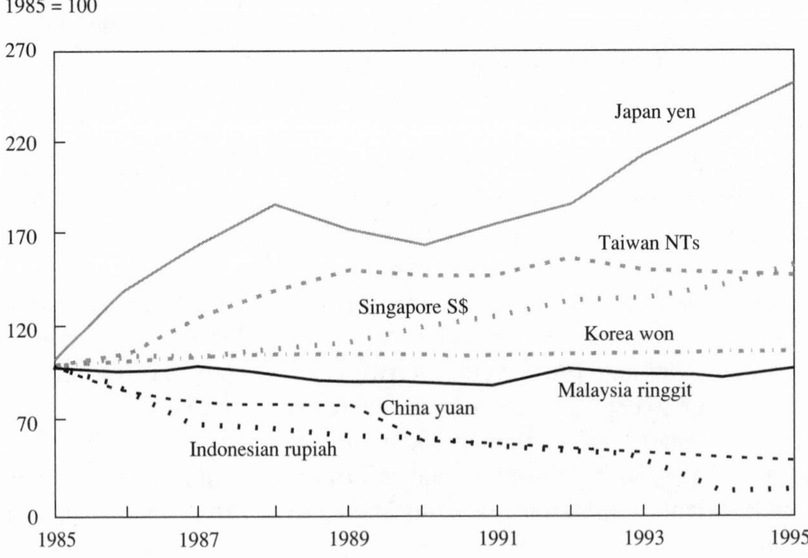

Source: International Monetary Fund, *International Financial Statistics* (Washington, D.C.: IMF, various years).

Table 4-1. *Foreign Direct Investment in Selected Asian Countries, 1986–95*
Millions of U.S. dollars

				China	
Year	*Malaysia*	*Thailand*	*Indonesia*	*Contracted*	*Actual*
1986	525	579	800	3,330	1,874
1987	750	1,949	1,240	4,319	2,314
1988	2,011	6,249	4,409	6,191	3,194
1989	3,401	7,995	4,719	6,294	3,392
1990	6,228	14,128	8,751	6,986	3,487
1991	5,554	4,988	8,778	12,422	4,366
1992	7,036	10,792	10,323	58,736	11,007
1993	2,297	4,294	8,144	111,435	27,514
1994	4,277	5,950	23,724	81,406	33,787
1995	3,660	16,436	39,915	90,288	37,736
1996	6,800	15,200	29,900	73,200	40,100

Sources: Malaysian Industrial Development Authority, Malaysia; Board of Investment, Thailand; Capital Investment Coordinating Board (BKPM), Indonesia and *China Statistical Yearbook* (Beijing: Zhongguo Tongji, various years). 1996 data for Thailand are preliminary. Data for Malaysia, Thailand, and Indonesia are approvals, not realized investment.

and Singapore, having displaced the United States and Europe. Granting the usual caveats on comparability of data from different-country sources, table 4-2 shows that Hong Kong was the largest investor in Thailand, accounting for 50.7 percent of the world total, followed by Japan (19.2 percent). In Malaysia, the top investor in 1990 was Taiwan (36.3 percent), followed by Japan (23.9 percent), while in Indonesia, Japan (25.6 percent) was the largest investor, followed by Hong Kong (11.4 percent). The NIEs as a group constituted the largest investors in the three major ASEAN countries. In all these countries, investments by the NIEs accounted for the lion's share, being 62.2 percent in Thailand, 47.2 percent in Malaysia, and 29.8 percent in Indonesia.

In Malaysia, the bulk of these Northeast Asian foreign investments were concentrated in electrical and electronics products, chemicals and chemical products, food manufacturing, textiles and textile products, wood and wood products, and basic metal products. In Indonesia, the concentration was in chemicals, paper and paper products, textiles, and metal products. In Thailand, most FDI was in electrical and electronic products, chemicals, textiles, and machinery and transport equipment.

This surge of foreign investment inflows from Northeast Asia into ASEAN, together with the dynamic effect of the liberalizing and dereg-

Table 4-2. *Percentage Distribution of Investments from Asia, Europe, and the United States in Three ASEAN Countries, 1990*

From	*Thailand*	*Malaysia*	*Indonesia*
Asia	81.4	85.1	57.0
Japan	19.2	23.9	25.6
Newly industrializing economies	62.2	47.2	29.8
Hong Kong	50.7	2.1	11.4
South Korea	1.9	3.7	8.3
Singapore	4.2	5.1	3.0
Taiwan	5.4	36.3	7.1
Europe	10.3	8.3	12.2
United States	7.7	3.2	1.8
World total (percent)	100.0	100.0	100.0
World total (millions of U.S. dollars)	14,128	6,228	8,750

Sources: Board of Investment, Thailand; Indonesia Investment Coordinating Board; and Malaysian Industrial Development Authority.

ulatory measures undertaken domestically, resulted in an ASEAN economic boom between 1987 and 1995. Annual real GDP growth during this period was 8.9 percent for Singapore, 9.5 percent for Thailand, 8.5 percent for Malaysia, and 6.8 percent for Indonesia, significantly higher than the 5.5 percent, 5.4 percent, 4.5 percent, and 4.9 percent achieved, respectively, during 1981–86 (table 4-3).

In the meantime, in a sharp break with previous policies, China in 1978 began to actively pursue foreign capital, as part of the program for opening up the economy. Such investment was also viewed as an effective way to transfer technology and managerial expertise to China. In 1978 it was announced that China would welcome foreign direct investment, and legislation governing such investment was adopted in the following year. In line with this approach, special economic zones were established in 1980 as the principal areas for direct investment. SEZs were set up along the southeastern coast of China at Shenzhen (near Hong Kong), Shantou (north of Hong Kong), Zhuhai (near Macao), and Xiamen (near Taiwan), positioned to attract investments from overseas Chinese. One crucial difference between SEZs and other areas in China was the administrative decentralization that permitted SEZ authorities to operate largely outside the state plan. They were allowed to attract foreign investors through preferential policies, undertaking their own infrastructure development by raising their own funds.

Table 4-3. *Growth Rate of Gross Domestic Product in Five ASEAN Countries, 1981–96*
Percent

Year	*Singapore*	*Malaysia*	*Thailand*	*Indonesia*	*Philippines*
1981	9.6	6.9	6.3	7.9	3.9
1982	6.9	5.9	4.1	2.2	2.9
1983	8.2	6.3	7.3	4.2	0.9
1984	8.3	7.8	7.1	6.7	–6.0
1985	-1.6	-1.0	3.5	2.5	–4.3
1986	1.8	1.2	4.5	5.9	1.4
1981–86	5.5	4.5	5.4	4.9	–0.2
1987	9.4	5.2	9.5	4.9	4.7
1988	11.1	8.9	13.2	5.7	6.3
1989	9.2	8.8	12.0	7.4	5.6
1990	8.3	9.8	10.0	7.2	2.7
1991	6.7	8.6	8.2	7.0	–0.6
1992	6.0	7.8	7.4	6.5	0.3
1993	10.4	8.3	8.2	6.5	2.1
1994	10.2	9.2	8.8	7.5	4.4
1995	8.9	9.5	8.5	8.1	4.5
1987–95	8.9	8.5	9.5	6.8	3.3
1996[a]	7.0	8.2	6.7	7.8	5.5

Sources: *Asian Development Outlook* (Oxford: Oxford University Press, 1996).

a. 1996 data are projections from applicable *Country Reports* (London: Economic Intelligence Unit, 1997).

In 1984 it was decided to promote foreign investment on a wider geographic basis. To this end, fourteen coastal cities and Hainan Island were permitted to offer tax incentives for such investment similar to those offered by the SEZs. In these areas imports by foreign investors of machinery, equipment, and other inputs were exempt from import licenses and customs duties, and a special 15 percent preferential income tax was applied to foreign investment enterprises. The extensive liberalization and the opening of the fourteen coastal cities led to an upsurge of FDI in 1985, reaching an unprecedented contracted amount of $6.3 billion.

Although China had considerable success in attracting foreign investment, objectives in that area were not completely fulfilled between 1980 and 1990. There were increasing complaints from foreign investors, particularly non-Chinese foreign investors, that the business environment in China was not favorable. In particular, they cited the artificially high costs of key inputs (land, office space, and labor), inadequacies of infrastructure (especially transportation and energy), unclear rules, punitive charges, and problems in dealing with the bureaucracy. More significant, the initial requirement that foreign investment enterprises had to balance their foreign exchange earnings was a major impediment to the expansion of foreign investment, in part because this requirement made it difficult for enterprises to repatriate profits. It also severely restrained investment in projects designed to produce for the domestic market, the key attraction in investing in China for non-overseas-Chinese investors. These problems, as well as the cyclical swing in the macroeconomic cycle, resulted in the contracted amount of FDI declining to about \$3–\$4 billion between 1986 and 1987; ASEAN countries attracted the bulk of the outflows of FDIs from Japan and the NIEs.

To address the problems that emerged, the Chinese State Council issued draft regulations on direct investment in October 1986. The regulations provided for reductions in land-use fees, taxes, costs of certain inputs, and labor costs. Improved access was promised for important inputs under state control, particularly for transportation and energy. Approval and licensing procedures for foreign investment enterprises were streamlined, and greater autonomy of these enterprises over production plans, imports and exports, wages, and terms of employment was guaranteed. In addition, to address the problem of balancing foreign exchange accounts, the State Council established foreign exchange adjustment centers. In 1988 local authorities in a number of areas created one-stop offices to speed up the processing of permits needed by foreign investment enterprises to begin or expand operations. Direct investment picked up to about \$6 billion in 1988-89 but stagnated in 1990 because of the violence in Tiananmen Square, where demonstrators were crushed by the military (table 4-1).

Tiananmen introduced considerable uncertainty into the business and investment climate in China. Consequently, in 1989 and 1990 outflows of investments from Japan and the NIEs were again largely diverted from China to ASEAN. Whereas FDI stagnated at \$6.9 billion in China in 1990, it doubled to a total of \$29.3 billion in Malaysia, Indonesia, and Thailand (table 4-1).

Challenge from China, 1991–93

In May 1990 the State Council announced the opening and development of the Pudong New Area, a zone of 135 square miles adjacent to the City of Shanghai, as a new economic and technological development area. According to the Shanghai City government's plan, Pudong New Area will be developed in several stages spanning thirty to forty years, with the first and second stages covering the eighth and ninth five-year-plan periods. It is envisaged that the project's multibillion dollar infrastructure requirements will be financed by the central and municipal governments, domestic bank loans, and foreign capital. The State Council and municipal government released regulations that granted foreign investors in Pudong New Area preferential treatment similar to that granted in special economic zones.

In the latter part of the 1980s several of the coastal cities developed a special system for the paid transfer of the right to use land to encourage foreign investors to plan long-term investment. In May 1990 the State Council issued general regulations for the sale and transfer of land-use rights in cities and towns. Under these regulations, companies, enterprises, organizations, and individuals were permitted to obtain land-use rights and undertake land development. Land users obtaining land-use rights could resell, rent out, mortgage, or use such rights during the validity period of the rights. The maximum period for land-use rights ranged from forty years for commercial, touristic, or recreational uses to fifty years for industrial use and seventy years for residential use.

In late January 1992 Deng Xiaoping made a historic tour of south China, where he made a speech calling for faster economic growth and deeper economic reform. This speech was later circulated as Internal Party Document 2 for senior cadres to "study." This call for greater economic reforms sparked off an unprecedented investment boom in south China, which soon spread to other areas. Some of the investments came because China rapidly relaxed many restrictions on foreign investors in early 1992. Foreign-invested factories for consumer products, which were formerly required to export the bulk of their output, were now allowed to target China's 1.16 billion consumers. Meanwhile foreign banks and law firms were getting more leeway to operate, foreign capital was encouraged to enter infrastructural development, and foreign property developers were encouraged to build apartments, hotels, office towers, and shopping malls. Local authorities were also given greater power to accept foreign investment without Beijing's approval. FDI contracted (which almost doubled from $6.6 billion in 1990 to $12.0 billion in 1991) skyrocketed to $58.7 billion in 1992 and $111.4 billion in 1993.[7] Foreign capital utilized also rose substantially, from $3.5 billion in 1990 to $4.4 billion in 1991 and more than doubled again to $11.0 billion in 1992. The momentum continued unabated, to reach $28 billion in 1993 and $40 billion in 1996 (table 4-1).

This upsurge of FDI in China since early 1992 was increasingly at the expense of ASEAN. The surge of Northeast Asian investment into Southeast Asia between 1986 and 1990 moderated. While the ASEAN big three (Indonesia, Malaysia, and Thailand) managed to capture 80 percent of total contracted FDI in Southeast Asia and China in 1990, their share declined steadily to 59 percent in 1991. By 1992 it had fallen to only 40 percent, with China accounting for 58 percent and Vietnam's share rising to about 3 percent.

More important, the decline of FDI in ASEAN would have been more severe if the 1992 investments contracted did not include several large petroleum and mining projects. In Malaysia, a petroleum-refining facility proposed by Société National Elf Aquitane of France accounted for M$4.1 billion, or about a quarter of the total foreign investments in 1992. The bulk of U.S. investment of M$3.3 billion was also petroleum related. Approved investments from Japan actually fell 30.6 percent, from M$3.8 billion in 1991 to M$2.6 billion, and those from Taiwan by 58.3 percent, from M$3.6 billion to M$1.5 billion. Excluding petroleum, foreign direct investment in the manufacturing sector in Malaysia is estimated to have fallen from M$13.4 billion in 1991 to M$5.4 billion in 1992.

More disturbing, domestic investments by Malaysians also fell 33.3 percent, from M$14.5 billion in 1991 to M$9.7 billion in 1992. Moreover, investment applications, an indication of future FDI in the pipeline, also decreased 42 percent, from M$20.2 billion (901 applications) in 1991 to M$11.8 billion (863 applications) in 1992. Although this substantial decline could be partly due to Malaysian authorities becoming more selective and shifting their focus to quality projects that emphasize higher levels of technology, skill, and capital, it could also signal a shift in Malaysia's competitive position in the scramble for foreign capital.

In Indonesia, the increase in FDI from $8.8 billion in 1991 to $10.2 billion in 1992 was attributed to several large petroleum and mining projects. These included the $1.5 billion Tanjung Uban Exor Refinery, the $1.6 billion Chandra Asri Olefin plant in West Java, and the $0.8 billion expansion of Freeport copper mining in Irian Jaya. Professor Soewito of Diponegoro University in Semarang, Central Java, estimated that foreign investment, excluding petroleum and mining projects, plunged by 54 percent in 1992. More significant, domestic investment declined 34.3 percent, from 41.1 trillion rupiah in 1991 to 27.0 trillion rupiah in 1992. These substantial downturns in domestic and foreign investments aroused considerable concern in Indonesia. A legislator of the Indonesia Democratic party, Aberson Marle Sihaloho, urged the government to monitor and control Indonesian investment in China to prevent the depletion of the nation's wealth.[8] Sihaloho noted that wealthy Indonesian Chinese businesses acquired their fortunes in Indonesia thanks to the facilities and assistance given by the government.

In Thailand, the number of applications from potential foreign investors seeking Board of Investment (BOI) privileges also declined during the first ten months of 1992. From January to October 1992, 369 projects were submitted to the BOI, a decline of 166 projects (31 percent) from the 535 recorded during the corresponding period in 1991. The combined investment capital of proposals dropped 24 percent, from 204,798 million baht to 155,900 million baht. The biggest decline was in the number of projects proposed by investors from Japan: the number of applications fell 45 percent, from 112 projects to 62 projects. The second largest decline was in investors from Europe, who submitted only 45 projects during the period, down 29 percent. The number of applications from Taiwanese investors also dropped 13 percent. U.S. investors submitted 29 projects, down 37 percent from the 46 projects in 1991, while the number of applications from Hong Kong investors dropped 27 percent. This decline alarmed Thai authorities. Officials from six overseas investment offices were in Bangkok in January 1993 for a meeting held by the BOI to discuss "offensive" strategies to promote foreign investment in Thailand. BOI secretary-general Staporn Kavitanon reportedly met the chiefs of Economic Counselor's Offices in Frankfurt, New York, Sydney, Tokyo, Hong Kong, and Paris to brief them on the new strategies to attract foreign investors.

Response from Southeast Asia, 1993–96

The competitive pressure from China powerfully affected the process of trade and investment liberalization in Southeast Asia, particularly Indonesia. The various reform packages announced by the

Indonesian government since 1992 were responding to Chinese competition as well as to broader trends in regional and global competition. The main objectives were to further ease foreign investment requirements, reduce import protection, ease bureaucratic impediments, and streamline the duty drawback scheme for exporters.

An important step in deregulation was taken in July 1992, when 100 percent foreign ownership was permitted in the following three types of investment: projects worth at least $50 million; projects located in any one of fourteen less-developed provinces (primarily in eastern Indonesia); and projects worth less than $50 million and located outside any of the fourteen provinces but situated in bonded zones, with 100 percent of production to be exported.

A June 1993 reform package further reduced the number of sectors in the negative investment list, from fifty-one to thirty-three. Among the sectors opened to foreign investors were nonautomotive internal combustion piston engine manufacturing; heavy equipment manufacturing, including bulldozers, loaders, graders, excavators, road rollers, and diesel forklift trucks; three-wheeled vehicles; the breeding of certain chicken stocks; fruit wines; and wheat flour.

In October 1993 the Indonesian government again further addressed the principal concerns of foreign investors: the bottlenecks in infrastructure (physical infrastructure, electricity, and communication), long and complex licensing procedures, complications in obtaining land titles, ownership limitations, and the lack of transparent rules for business dealings. Reflecting Indonesia's concerns about competitiveness in attracting foreign investment, the deregulation package focused on easing investment licensing restrictions at the provincial level, relaxing divestiture requirements, lowering minimum capital requirements for some investments, and streamlining procedures for environmental impact assessments. One of the most important measures entailed simplifying investment licensing procedures by eliminating some of the layers of bureaucracy. The deregulation package also further relaxed ownership restrictions by lowering the investment threshold for 100 percent foreign-owned companies, from $50 million to $4 million, provided that these companies produce inputs or components for other industries.

The October 1993 package also extended the time span during which foreign investors were allowed to fully own industries in Indonesia. The divestment process must start in the eleventh year of commercial production, as against the sixth year previously. The divestment of shares, which should take place within twenty years, can be carried out through the capital market or direct placement rather than through a forced sale to designated local partners.

These liberalization reform measures, particularly the easing of restrictions on foreign ownership and forced divestment, improved the competitiveness of Indonesia in attracting FDI compared with China, Vietnam, and South Asia. Whereas FDI fell from $10.3 billion in 1992 to $8.1 billion in 1993, with a particularly severe decline from Japan, Hong Kong, Taiwan, and the United States, FDI rose to a record $23.7 billion in 1994. Major infrastructure projects like power plants and refineries substantially boosted the approved FDI.

To sustain its attractiveness to foreign capital compared with China, Indonesia embarked on another round of investment and trade liberalization in 1994. The principal features included the lifting of the ban on foreign investment in hitherto strategic sectors like ports, telecommunications, power, railway, and civil aviation; the easing of equity limits on foreign partners in joint ventures, to 95 percent from 80 percent; the approving of 100 percent foreign-owned enterprises for the whole of Indonesia instead of merely in remote areas; the extending of the validity of foreign companies' business licenses from twenty to thirty years; the further easing of divestment requirements, with only a token divestment of 1 percent required after fifteen years; and the easing of operating conditions for foreign companies, particularly the permission to establish subsidiaries and acquire the foreign and domestic firms.

These extensive liberalization measures, particularly for infrastructural investment, led to a large increase in investment. Approved FDI rose from $23.7 billion in 1994 to $39.9 billion in 1995. Approved FDI dropped somewhat in 1996 but remained robust at $29.9 billion (table 4-1).

On closer scrutiny, however, the approved FDI for 1995 included thirteen oil refineries, some of which were refinery projects owned by Indonesians. If one were to exclude the $19.4 billion in chemicals, foreign direct investment in employment-generating sectors like food, textiles, wood, paper, metal goods, and nonmetallic minerals actually declined from $8.9 billion in 1994 to $7.2 billion in 1995, indicating the continued competitive pressure on Indonesia in attracting FDI.

For Malaysia, the decline in FDI from $7.0 billion in 1992 to $2.3 billion in 1993 was a shock. Though the momentum of projects in the pipeline could carry the economy over the next several years, sustained weakness in FDI would result in reduced manufacturing and overall economic growth. Consequently, further trade and investment liberalization as well as the domestic investment initiative was started in 1993 to boost growth. In 1994 FDI rose to $4.3 billion, almost double the amount for the previous year. Significantly, the southward strategy of Taiwan again made it the largest investor in Malaysia (24 percent), with other Asian countries such as Japan (20 percent), Singapore (9 percent), and Hong Kong (8 percent) dominating as a group. The U.S. share moderated to 11 percent.

Recognizing the relentless competitive pressure from China, Vietnam, India and other developing countries, Malaysia made a concerted effort to upgrade its industrial structure. High-tech industrial parks were developed in Kulim of Kedah, where companies like Intel produced high-end motherboards and networking-communications cards. Employment rules, particularly on the hiring of foreign skilled engineers, were significantly relaxed. In 1996 Malaysia launched its cluster promotion strategy, patterned on Singapore's 1992 strategic economic plan. Aside from resource-based clusters like petrochemicals and chemical products, promotion was focused on key industries like electronics, aerospace, automotive, and machinery and equipment. Public-sector support for research and development received a major boost of M$1 billion Malaysian to complement the upgrading effort. FDI approvals rebounded strongly in 1996, to $3.66 billion (table 4-1).

In Thailand, the significant decline in FDI approvals in 1993 and 1994 pointed to signs of waning interest in the country's attractiveness as a manufacturing center. Disadvantages included poor infrastructure, excessive red tape, and a manpower shortage. The Board of Investment, acknowledging the pressure to upgrade, has been actively discouraging capital inflows into industries like textiles and shoes while courting sectors with higher-skill and value-added potential, to facilitate the economy's transition from an assembly base to a total manufacturing center. The focus has been on building on Thailand's existing strength in motor vehicles, petrochemicals, steelmaking, and electronics, with the emphasis on developing a cluster of core industries and component suppliers. The strategy was designed to enhance Thailand as a regional base of multinationals seeking a springboard into the ASEAN Free Trade Area (AFTA) before that preferential trading arrangement comes into full force in 2003. Those firms inside AFTA

Table 4-4. *Accelerated Tariff Reduction Schedule for ASEAN Free Trade Area*

Track	*Initial timetable (Singapore Summit, January 1992)*	*Accelerated timetable (Chiangmai Meeting, September 1994)*
Normal		
Tariffs > 20 percent	20 percent by 2001 15 percent by 2003 10 percent by 2005 0–5 percent by 2008	20 percent by 1998 0–5 percent by 2003
Tariffs < 20 percent	15 percent by 2003 10 percent by 2005 0–5 percent by 2008	0–5 percent by 2000
Fast		
Tariffs > 20 percent	0–5 percent by 2003	0–5 percent by 2000
Tariffs < 20 percent	0–5 percent by 2000	0–5 percent by 1998

Source: ASEAN secretariat.

would then have an advantage over those outside in tapping into the consumer markets of more than 400 million people.

Apart from individual responses at the country level, ASEAN countries collectively responded to the competitive pressure from China as well as to broader regional and global competitive trends by accelerating the liberalization program in AFTA. Initially, the Singapore summit held in January 1992 endorsed the Thai proposal of establishing AFTA within fifteen years after January 1, 1993, using the common effective preferential tariff scheme (CEPT). Under this scheme, tariffs for fifteen categories of products currently above 20 percent are to be reduced to 20 percent within five to eight years. Subsequent tariff reduction from the 20 percent level to 0–5 percent are to be spread over a maximum of seven years. The goal was for tariffs to reach the 0–5 percent mark in fifteen years for all manufactured goods, processed agricultural products, and capital goods (table 4-4).

Meeting in Manila in October 1992, the ASEAN economic ministers decided that the tariffs on those items of more than 20 percent were to be lowered to a maximum of 5 percent within ten years, while products with tariffs that were then 20 percent or less would be reduced to no more than 5 percent within seven years. Because of pressure from the forward movement of APEC (Asia-Pacific Economic Cooperation), the completion of the Uruguay Round negotiation, and the ratification of the World Trade Organization, the original AFTA agreement in 1992 again had its timetable accelerated and its scope broadened and deepened. More specifically, at the ASEAN economic ministers' meeting in Chiangmai in September 1994, the timetable for all CEPT products to lower tariff rates to 0–5 percent was accelerated from fifteen to ten years (that is, by 2003 instead of 2008). Thus the target date for AFTA was brought forward to 2003 (table 4-4). The product coverage of CEPT was also broadened to include unprocessed agricultural products. More important, discipline was introduced in the use of exclusion lists by requiring that products in the temporary exclusion list be transferred to the inclusion list in five equal installments of 20 percent, beginning January 1, 1996.

Even more significant, the competitive response of Southeast Asia to the challenge of China created a unique window of opportunity for the Taiwanese government to use its economic and capital muscle to counter China's increasing embrace through a southward strategy. This strategy was conceived in early 1993 to diversify Taiwan's investment linkages and economic dependence away from China toward Southeast Asia and to improve the political room for maneuver. The objective was to use party and government enterprises to develop industrial estates in Southeast Asia, supplemented by an investment guarantee agreement and a treaty to avoid double taxation, to lure Taiwanese investors from China into Southeast Asia. In addition, the Taiwanese government would provide soft loans to Southeast Asian governments to finance the development of industrial parks, export processing zones, and other infrastructural projects that catered to the needs of Taiwanese investors. Through this strategy Taiwan was also looking forward to the AFTA as an expanding market to counter the attraction of the China market for Taiwanese products in the long run.

Preliminary data on Taiwan's investment in Southeast Asia shows that the southward strategy has had some success. For the whole of 1994 Taiwanese investment in Southeast Asia and Vietnam surged by 320 percent, to reach $4.79 billion. At the same time contracted investment into China fell by 46 percent, to reach $3.39 billion.[9] Large Taiwanese business groups that have ventured into the Southeast Asian region include China General Plastics, with a $480 million petrochemical project in Malaysia; Far Eastern Textile, with synthetic fibers production facilities in the Philippines; Hualon, with production base in

Malaysia; Tuntex in Thailand; Chung Hwa Pulp in Indonesia; and Acer Computer in Penang and Singapore.

By early 1995 the largest projects inspired by the ruling party (the Kuomintang, or KMT) were in Indonesia. The China Development Corporation, headed by Liu Taiying, chairman of the KMT's Business Management Committee and a former classmate of President Lee Teng Hui at Cornell University in the 1960s, entered into partnership with the Djajanti Group. Their joint ventures included a $600 million cement plant in eastern Indonesia as well as a $1 billion petrochemical plant near Surabaya.

In Vietnam, the development of an industrial zone outside Ho Chi Minh City was undertaken by the Central Trading and Development Group (CTD), a trading company owned by the KMT. The CTD has committed itself to four projects with total approved investments of $600 million. These include the development and management of the 300-hectare Tan Thuan Export Processing Zone and a 675-megawatt power plant south of the city. In addition, the CTD has teamed up with Ho Chi Minh City to develop a 2,600-hectare development zone called Saigon South, which is to include universities, hotels, parks, and sports facilities, as well as commercial and residential zones.

The number of such Taiwanese infrastructural and chemical projects has continued to grow, particularly in Indonesia. In 1996 Chinese Petroleum of Taiwan, the state-owned firm, signed a memorandum of understanding to transfer a naphtha cracker facility to Indonesia. The plant, with an investment of $2 billion, would produce 230,000 metric tons of ethylene annually. Seven Taiwanese petrochemical firms would build auxiliary plants in partnership with Indonesian chemical companies. Moreover, the Yieh Phui Enterprise of Taiwan would invest $750 million in a steel factory with the Lippo Group of Indonesia.

Conclusion

The competition for scarce capital—particularly in the form of foreign direct investment, with its employment creation, wage increase, transfer of industrial technology, managerial expertise, marketing know-how, and stimulus to the development of local supporting and domestic industries—is likely to intensify in East Asia. That will be especially true as the political legitimacy of both democratic and authoritarian governments in the region becomes increasingly dependent on delivering goods to their people, who are increasingly able to compare their living standards with those of neighboring countries. This competitive pressure is a key factor driving the process of unilateral liberalization in the trading and investment regime in the region.

It is instructive to note that the competitive pressure of foreign corporations on local businesses in East Asia in the past decade did not result in a strict protectionist policy on foreign capital, partly because of relatively strong autonomous states able to fend off sectional interests and pressure. Moreover, the significant effect of employment and wage increases on the general population, the beneficial effect of FDI on local partners with political connections, and the multiplier effect of FDI on local supporting industries weakened the resistance to foreign capital in East Asian countries. These factors were also instrumental in spearheading the competitive, unilateral liberalization policies on trade and investment undertaken by most of these countries in the past decade. That laid the foundation for the decision at the November 1994 APEC summit in Bogor, Indonesia, to create a free trade area in the Asia-Pacific region by 2020.

From 1986 to 1990, as the ASEAN countries rapidly liberalized and boomed, China was still mired in bureaucratic red tape, poor infrastructure, and unclear legal and commercial rules, as well as being saddled by Tiananmen. Consequently, ASEAN countries garnered a huge share of FDI flowing out of Northeast Asia. At its peak in 1990 ASEAN obtained about 80 percent of the total global FDI flowing into East Asia, with China capturing only 18.9 percent and Vietnam 1.6 percent. The swing toward China from the low point of 1990 was dramatic, particularly in 1992 and 1993, when China's share of East Asian FDI rose to 58–80 percent, as against 20–40 percent for ASEAN.[10] Although the pendulum has since swung back in the direction of ASEAN, the rapidly developing Northeast Asian free trade area, spurred on by market forces that are weaving together the highly complementary economies of Japan, Korea, Taiwan, Hong Kong, and China, is posing serious challenges to AFTA.

The nature of the challenge from China differs for the various Southeast Asian economies. If the center of growth continues to shift from Southeast to Northeast Asia, Singapore, serving the Southeast Asian hinterland, could lag behind Hong Kong as the preeminent regional operating center. The diversion of FDIs from ASEAN contries toward

China has not yet significantly affected Singapore directly, since its per capita income ($24,000) and levels of wages, skills, and technology are still substantially higher than China's. Like the other NIEs, whose per capita income is $8,000–$23,000, Singapore is more complementary than competitive with the rapidly expanding Chinese economy, though segments of the rapidly developing cities and coastal region in China have begun to compete with the NIEs. However, with GDP per capita comparable to that of China ($370–$2,000, depending on the region or statistical methodology used), Indonesia ($700), Thailand ($2,000), and Malaysia ($4,000) are at wage, skill, and technology levels that are more competitive with various regions in China. The diversion of Japanese, Western, Taiwanese, and Hong Kong capital to China therefore had more severe effects on them. In addition, overseas-Chinese capital in these ASEAN countries also flooded into China, largely through companies located in Hong Kong. The simultaneous decline in FDI and local investment as a result of competitive pressure from China generated uneasiness, particularly in Indonesia.

More important, the substantial surplus labor to be released from the agricultural sector in China over the next two decades will probably sustain China's comparative advantage in labor-intensive products for much longer than Japan or the other NIEs sustained their advantage during their earlier period of labor-intensive industrialization. This might mean that real wages for unskilled and semiskilled workers in East Asia will remain depressed for a long time. As real wages increase in the coastal region in China, labor-intensive industries will migrate inland rather than to Southeast Asia, just as they have migrated from Japan to NIEs to ASEAN countries over the past twenty years. For a country at a similar stage of development with surplus labor (such as Indonesia), it might mean real living standards will remain depressed for a prolonged period. Income inequality as well as regional disparity could also worsen. That could have significant implications for social and political stability as well as ethnic Chinese and indigenous Indonesian relations. But this competitive dynamic also leaves considerable room for a capital-exporting economy like Taiwan to exploit as it formulates its southward strategy and vacation diplomacy.

The competitive industrial and trade structure between China and Southeast Asian countries such as Indonesia, Vietnam, Thailand, and to a lesser extent Malaysia, as well as Japan's strategy of integrating Southeast Asia into its regional industrial division of labor, could lead to an intensifying Japan-China rivalry playing out in Southeast Asia. Japan is interested in maintaining its head-goose position in the harmonious flying pattern, with Southeast Asian countries like Thailand, Vietnam, Indonesia, and Malaysia subsumed under an informal regional trade and investment grouping. The relocation of consumer electronics and automotive industries from Japan to Southeast Asia indicated this underlying strategy and the leveraging of the Southeast Asian industrial structure and expanding market to help Japan sustain its global competitiveness. But China is bound to contest this pyramidal industrial structure, and it may use the powerful and pervasive overseas-Chinese business groups in Southeast Asian countries to integrate these economies into the expanding China market so as to moderate Japanese influence. The active cultivation of Malaysian, Thai, and Singapore political and business groups by China is a sign of this emerging trend.

Notes

1. This is consistent with the lessons of endogenous growth theories, which stress that ideas are as important as an open economy. See Edward Y. K. Chen, *Multinational Corporations, Technology, and Employment* (St. Martin's Press, 1983); and Linda Y. C. Lim and E. F. Pang, *Foreign Investment and Industrialization in Malaysia, Singapore, Taiwan, and Thailand* (Paris: OECD Development Center, 1991).

2. Quoted in Chia Siow Yue, "Direct Foreign Investment and the Industrialization Process in Singapore," in Lim Chong Yah and Peter J. Lloyd, eds., *Singapore: Resources and Growth* (Oxford University Press, 1986), p. 102.

3. John Dunning, *International Production and the Multinational Enterprise* (London: Allen and Unwin, 1981).

4. Magnus Blomstrom, Robert E. Lipsey, and Mario Zejan, "What Explains Developing Country Growth?" Working Paper 4132 (Cambridge, Mass.: National Bureau of Economic Research, August 1992). (Hereafter NBER.)

5. Using city-level data for a sample of 434 Chinese cities during 1988–90, Shang-jin Wei found a significant effect of FDI on industrial output. A 1 percent increase in the size of FDI is associated with a 1.3 percent higher growth rate in industrial output, which supports the hypothesis of spillover of technological, managerial, and marketing know-how across firms within cities. See Wei Shang-jin, "Open Door Policy and China's Rapid Growth: Evidence from City-Level Data," Working Paper 4602 (NBER, December 1993). Based on a sample of 28 provinces, autonomous regions, and municipalities in China over the period 1981 to 1990, Ball, Khor, and Kochhar found that FDI (measured by the ratio of foreign capital actually utilized to total fixed assets investment) was a significant variable explaining the real growth of provincial per capita industrial output. Michael W. Ball, Hoe E. Khor, and Kalpana Kochhar, *China at the*

Threshold of a Market Economy (Washington: International Monetary Fund, September 1993).

6. There are as yet no comprehensive data on FDI stocks and flows in each of the ASEAN countries. FDI data usually come from the balance-of-payments statistics (International Financial Statistics, IMF) and from the investment boards. In Indonesia, the sources are Bank Indonesia and the Capital Investment Coordinating Board (BKPM); in Malaysia, Bank Negara, Department of Statistics and Malaysian Industrial Development Authority; in the Philippines, Central Bank of Philippines and Board of Investment; in Singapore, the Economic Development Board and the Department of Statistics; and in Thailand, Bank of Thailand and the Board of Investment. Because of substantial differences among national data in definition and coverage, the data cannot be aggregated to produce an accurate ASEAN total and have limitations for intercountry comparisons. For example, BKPM data on FDI in Indonesia exclude investments in banking and petroleum and gas, and it is well known that U.S investors dominate FDI in those sectors. Moreover, paid-up capital and balance-of-payments capital flows in FDI understate the extent of resources and facilities owned and controlled by foreigners, because they neglect foreign participation not accompanied by financial inflows.

7. This large amount of investment contracted (as reported by China's Ministry of Foreign Trade and Economic Cooperation) could be an overestimation, since localities competed to announce large contracts and projects. Some of the so-called foreign investments were disguised Chinese capital from Hong Kong taking advantage of preferential policies. However, the significant diversion effect from ASEAN countries to China was unmistakable.

8. "Control on RI Investments in China Urged," *Jakarta Post,* December 3, 1992.

9. Reports by the Cabinet's Council for Economic Planning and Development, as quoted in *United Daily News,* Taiwan, and Xinhua News Agency, April 7, 1995.

10. Of course, the significant upsurge of FDI into China also generated problems, including an expanding money supply and the ensuing inflation.

PART II

CHAPTER FIVE

Left for Dead: Asian Production Networks and the Revival of U.S. Electronics

Michael Borrus

FROM THE EARLY 1970s until the mid-1980s Japanese producers were ascendant in electronics. In short order, they had taken over consumer electronics, gained leading world market shares in semiconductor chips, materials, and equipment, and were looking capable of repeating the feat in computers, office systems (such as copiers and faxes), and customer telecommunications equipment. So worried were U.S. policymakers and industrialists that the avowedly laissez-faire Reagan administration took the unprecedented step of using interventionist industrial policy to support the domestic microelectronics industry.[1] If the rapid rates of attrition of U.S. market share had continued, U.S. firms would have joined their European counterparts as significant players only in niches and on the margin of mass global markets.

What a difference a decade made. By 1994 U.S. producers of silicon chips and semiconductor materials and equipment were again flourishing, having regained the dominant world position. U.S. producers of office, communications, and computer systems had reasserted product

This chapter is drawn from a larger work in progress on global competition in electronics. See Michael Borrus, *Punctuated Equilibria in Electronics: Microsystems, Standards' Competitions, and Asian Production Networks* (forthcoming, 1997).

and technical leadership, retaining clear market dominance, especially with the latter. As computer technology began to pervade consumer electronics, those same producers even looked to be reviving defunct U.S. consumer fortunes. By contrast, with few exceptions their once formidable Japanese competition appeared disorganized, dismayed, and decidedly on the defensive. Indeed, U.S. industry leaders are now so certain of continued success that many dismiss the Japanese giants as competitive dinosaurs, ill-adapted to the raucous, fast, changeable, idea-intensive electronics markets of the future.[2]

But it would be imprudent to conclude that U.S. firms are organizationally or strategically better placed than other competitors just because they thrived in the latest round of competition. Their success is the phenomenon to be explained, not the proof that those who thrived are better adapted for the future than those who did not.[3] Indeed, as argued later, the recent success of U.S.-owned firms has rested in large part on the growing technical sophistication and competitive strength of Asian-based producers in the China Circle, Singapore, and South Korea. Though useful to U.S. firms in the last round of market battles with Japanese firms, Korean electronics producers like Samsung and China Circle producers like Taiwan's dominant microcomputer firm, Acer, are also formidable potential competitors. As the Asia market develops in both technical sophistication and size over the next decades, the mantle of electronics leadership could well pass from U.S. and Japanese firms to indigenous Asian producers, especially those centered in the China Circle.

Any interpretation of the potential of the China Circle in electronics requires a clear understanding of the industry's recent competitive evolution, the principal focus here. Several competitive shifts lie behind recent Japanese troubles and American re-ascendance. Chief among these were the bursting of the domestic Japanese asset bubble, the attendant, lengthy recession in the Japanese economy, and multiple *endaka* (dramatic yen appreciation). Japan's past electronics success was in large part driven by rapid growth in the sheltered domestic market. Rapid domestic growth afforded the stable demand to reach scale economies, the launch market for several generations of consumer and office systems, premium prices to subsidize price competition on foreign markets, cheap capital for continuous reinvestment, and not least, quality- and feature-conscious consumers who rewarded corporate strategies built on incremental product revisions.[4] Cheap capital ended when the asset bubble burst, provoking Japan's longest postwar recession. Enduring recession put an end, at least temporarily, to the domestic economy's ability to support firm strategies premised on rapid growth and to the willingness of retailers to blindly support the producer-controlled pricing structure.[5] Combined with successive *endaka,* the economic problems made Japanese firms increasingly vulnerable to price competition both at home and abroad.

While recession is a temporary phenomenon, three simultaneous economic shocks are provoking enduring structural changes in strategy and behavior. Slower domestic growth for the foreseeable future and more costly capital have increased pressure on Japanese firms to scrutinize investment decisions more closely and to be more conscious of investment returns. Capital and technology are then likely to be turned over more slowly in production, moderating the ability of Japanese firms to compete through aggressive manufacturing innovation and the incremental product revisions it generates. The result will probably be not only increased specialization among Japanese electronics firms as they concentrate investment in areas of core advantage but also a thoroughgoing industry rationalization. Indeed, the rationalization is already visible in strains on the traditional employment and subcontractor systems and in the surge of offshore production investment in East Asia and China, as detailed by Dieter Ernst and Jean-François Huchet (chapters 7 and 8).[6]

Whereas Japan's economic problems begin to explain why its electronic giants faltered in the 1990s, they do not account for the resurgence of U.S. market and technical leadership. Two other competitive shifts are of paramount importance there—one in the market and one in production organization. The market shift was caused by a transformation in the nature of electronic systems products; as a result, the industry's principal business strategies changed dramatically. New electronics product markets have begun to converge on a common technological foundation of networkable, "open," microprocessor-based systems (of which, the personal computer [PC] is emblematic).[7] Such new product markets are characterized by a predominant form of market rivalry, namely, competitions to set de facto market standards—as Microsoft and Intel have done so successfully in PC operating systems and processor architectures, or as Cisco Systems has done with routers. Over the last half decade the domestic U.S. market has been the principal launch market for such new products and the principal terrain on which the resulting standards competitions have been fought. With just a few exceptions—for example, Nintendo in video games, Sony in 8mm video camcorders—U.S. firms have defined the products, set and controlled the standards, and consequently achieved dominant positions on world markets as U.S. choices became global standards.

The organizational shift was, however, just as important and in its own way permitted the new product-market strategies to succeed. The shift in U.S. firm production organization was the move away from traditional integration to network forms of organization—specifically, Asia-based production networks centered in the China Circle and Singapore.[8] The move to Asia-based production networks during the 1980s had three important consequences for U.S. firms. They were able to relieve the constraining threat of competitive dependence on Japanese firms for a wide range of component technologies and manufacturing capabilities, because their Asian production networks became a competitive supply-base alternative to Japanese producers.[9] Simultaneously, the networks helped to lower production costs and turnaround times while keeping pace with rapid technological progress—thereby permitting U.S. firms to pioneer strategies of continuous innovation.[10] Finally, the networks spawned Asian-based direct competitors to Japanese firms in several of their stronghold markets (such as memory chips, consumer electronics, and displays).

Home Base of Multinational Corporations

The rest of this chapter takes a closer look at the shift in production organization, the way it created an alternative supply base in Asia, and the role it played in the resurgence in U.S. firms to product and technical leadership in electronics. Before turning to those issues, however, it is appropriate to ask whether, in an industry dominated by multinational corporations (MNCs), an analysis that distinguishes between U.S.-, Japan-, and Asia-based industries still makes sense. The analysis here presumes that the international market dynamic in most high-tech industries can still be effectively analyzed as a competition between firms operating out of largely national home bases.[11] By "home base" I mean the national market in which most of a firm's assets, employment, and sales reside, and from which corporate control is exercised (especially control over strategy formation, corporate reorganization, new product development, finance, and distribution). In most cases the home base is also the predominant locus of corporate ownership.

By that definition, very few high-tech MNCs are globally footloose. Indeed, two-thirds to three-quarters of the assets, employment, and sales of most MNCs, and an overwhelming percentage of their best-compensated and highest-skilled jobs, are still in a home base.[12] Of the world's top fifty MNCs of all national origins, which might be expected to be the most nonnational of MNCs, almost all fall in the 60-90 percent range of assets within the home country.[13] Equally significant, almost all MNCs still explicitly exercise control from their home country of origin.[14]

Given those facts, my analysis here views firm strategies as systematically shaped by the logic of competition in the home market base. Domestic institutions shape a national market logic or system of production[15]—that is, characteristic ways of doing business and distinctive trajectories of technology development that are the basis of product differentiation on international markets.[16] For high-tech industries, the principal domestic institutional variables include (1) the structure of the industry in question and of its domestic market (such as oligopolistic, *keiretsu*, or lead customers); (2) technology, trade, and industrial policies and the political system that implements them; (3) the capital and labor market structures that condition access to those factor inputs; and (4) the local supply base that enables access to technology factor inputs.[17]

Those variables create a fabric of possibilities, a pattern of constraint and opportunity that confronts firms as they choose strategies, which make some choices more likely (or less risky) and foreclosing others. Consider, for example, how U.S. antitrust enforcement denies to U.S. firms the use of market-sharing arrangements that are routinely adopted in Japan and parts of Europe. Or consider how Japan's lifetime employment system encouraged corporate strategies built on in-house training and skill improvement of technical employees. Or how *guanxi* networks (based on frienship or kinship ties) permit smaller Taiwanese family "firms" to deal in high-risk international ventures.[18]

As such examples suggest, the home base's pattern of constraint and opportunity channels, in characteristic directions, corporate strategies and behavior and, through them, technology development. For instance, a well-developed venture capital market, highly flexible labor market, leading-edge military and computer industry demand, and competitive industry structure characterized by easy entry and exit, all shaped a U.S.-based semiconductor industry with distinctive strategies and technologies based on radical product innovation.[19] By contrast, *keiretsu*-dominated capital and distribution, inflexible labor markets, price-sensitive consumer demand, and a panoply of industrial and trade policies, all shaped a Japanese semiconductor industry with equally distinctive strategies and technologies based, in contrast to the U.S. pattern, on incremental manufacturing innovation.

Of course, strategies can and do differ among firms facing similar constraints, not least because they start with different resources and actively respond to what their competitors are doing. Firms can seek external opportunities or devise ways around national constraints. U.S. firms did exactly that by creating their Asia-based production networks. In the real world of commerce, then, the home-base institutions that shape a national system of production are systemic constraints tending to push strategies in particular directions.

The inherent openness of this analysis permits revision over time as evidence accrues to challenge the hypotheses it generates. Indeed, this chapter suggests that regional and subregional production systems in electronics may be gradually supplanting national ones. That would be an unintended consequence of the Asia-based production network strategy of U.S. firms, the subregional production networks it helped to spawn throughout Asia under the control of overseas-Chinese capital, and the parallel regional response of Japanese firms. If such developments were to much diminish the significance of the national home base, they would obviously require a revision of the approach adopted here.

Until then, however, my working hypothesis is that for most firms the national market logic dominates international market strategies. That holds true especially for the dominant Japanese electronics firms but also even for the U.S.-based MNCs that adjusted to high-tech competition by constructing production networks outside the United States. For the U.S. firms, in important ways, the home base has become more significant in the last ten years of increasing global competition than it had been earlier in the era of clearly defined national industries. As U.S. firms regrouped and restructured, they sought renewed competitive advantage from home-based sources. Indeed, the *global* leadership of U.S. firms was rebuilt partly on a *domestic* foundation—supportive policies, the U.S. market's tremendous competitive ferment, and its leadership both in the networking of microcomputer-based systems and in the design, product definition, and systems architecture capabilities that created the new standards.

U.S. Foreign Direct Investment and the Creation of a Regional Supply Base

By the end of the 1970s U.S. electronics firms were almost completely dependent on Japanese competitors for the supply of the underlying component technologies (such as tuners, picture tubes, recording heads, and miniature motors) necessary to produce consumer electronics products.[20] In most cases, thoroughgoing technology dependence was a first step toward market exit. It meant that U.S. firms were far enough removed from the technological state of the art to impede new product development and that their principal competitors could dictate time to market, product cost, and feature quality. Under those circumstances, profits were minimal—if any were to be had at all. Consequently, by 1980 most major U.S. firms had left the consumer segment of the market; remaining players like General Electric and RCA survived largely by putting their brands on Japanese original equipment manufacturing (OEM) production. A few years later, even RCA and GE, which had created most of the consumer electronics technologies that Japanese firms perfected, left the business.

The loss of the high-volume demand for consumer electronics eroded the U.S. *supply base* for the other segments of the electronics industry and threatened them with a constraining *architecture of supply.*[21] The supply base is the local capability to supply the component, machinery, materials, and control technologies (such as software), and the associated know-how, that producers use to develop and manufacture products. The architecture of supply is the structure of the markets and other organized interactions (such as joint development) through which the underlying technologies reach producers. In effect, U.S. producers of industrial electronics (such as computers and communications) were in danger of becoming dependent on their Japanese competitors for memory chips, displays, precision components, and a wealth of the other essential technologies (and associated manufacturing skills) that went into electronic systems.[22] The only alternative to increasing dependence on a closed oligopoly of rivals was to make the supply architecture more open and competitive. In conjunction with government policies and local private investors in Asia, U.S. firms gradually turned their Asian production networks into a flexible supply-base alternative to Japanese firms.

The transformation from cheap labor affiliates to alternative supply base occurred in three stages: an initial stage, from the late 1960s to the late 1970s, during which U.S. firms established their presence through foreign direct investments; a second stage, from 1980 to 1985, in which their Asian affiliates developed extensive local relationships in the shadow of the dollar appreciation; and a third stage, from the late 1980s to the early 1990s, when U.S. firms significantly upgraded the technical capabilities of their regional production networks and assigned local affiliates global product responsibilities. The U.S. progression from simple assembly affiliate to technologically able Asian production network contrasts sharply with the development pattern of Japanese investments in the region over the same period. A brief review of key developments in each of the three stages will highlight the differences.[23]

The First Stage

From the late 1960s, after an earlier round of market access investments by a few large U.S. multinationals (notably IBM, GE, and RCA), most U.S. firms sought not market access but cheap production locations in Asia. U.S. investment was led by U.S. chip makers, then consumer electronics and calculator producers, and finally, toward the end of the 1970s, producers of industrial electronic systems like computers and peripherals. Most of the U.S. investments in this first stage established local assembly affiliates. Cheap but disciplined Asian labor permitted U.S. firms to compete on price back home and in Europe. Right from the start, then, the Asian affiliates of U.S. electronics firms were established as part of a multinational production network to serve advanced country markets. By contrast, as Dieter Ernst suggests in his chapter, most Japanese investment in Asia during this period, led by consumer electronics and appliance makers, was aimed at serving nascent local markets behind tariff walls. Japanese investment was often turnkey, with knock-down kits exported from Japan for local final assembly and sale in the local affiliate's domestic market. Though the Japanese and U.S. investments in this first stage were both oriented to simple assembly and superficially appear similar, the vastly different markets being served pulled their respective investments in divergent directions.

Consider the resulting logic of sunk investment for the two sets of firms. Because their Asian affiliates were integrated into a production operation serving advanced-country markets, U.S. firms upgraded their Asian investments in line with the pace of development of the lead market being served, the U.S. market. In essence, they upgraded in line with U.S. rather than local product cycles. By contrast, Japanese firms were led to upgrade the technological capacities of their Asian investments only at the slower pace necessary to serve lagging local markets. As local U.S. affiliates became more sophisticated through several rounds of reinvestment, a division of labor premised on increasing local technical specialization developed throughout the U.S. firms' global production operations. Local needs began to diverge from those elsewhere in U.S. firms' total operations, and affiliates sought out, and where necessary trained, local partners to meet them.

To be sure, the growth of local autonomy and relationships was constrained by overall corporate strategies (for example, where economies of scale dictated a global rather than a local sourcing arrangement), but over time U.S. investments still led to greater technology transfer and increasing technological capabilities for locals. By contrast, stuck in developing-market product cycles, offshore Japanese affiliates benefited from no such incentives to upgrade and no need to develop local supply relationships. Japanese firms served the domestic and U.S. markets wholly from home. Whatever their lagging Asian affiliates needed could be easily supplied from Japan. As local Asian markets demanded the marginally more sophisticated goods whose product cycles had already peaked in the advanced countries, the entire production capability for them could also be transferred from Japan. Overall, less technology was transferred, and even that remained locked up within the Japanese firm's more limited circle of relations.

The Second Stage

During the second stage (1980–85), therefore, U.S.-owned assembly platforms were upgraded and enhanced technically to include more value-added; for example, from assembly to test in chips, from hand to automation assembly techniques, and from simple assembly of printed circuit boards to more complex subsystems and final assembly in industrial electronics. As they gained more autonomy, U.S. affiliates began to source more parts and components locally, including a range of mechanical parts, monitors, discrete chips, and even power supplies.

As U.S. affiliates developed and as the U.S. industry left the consumer segment, local electronics producers in places like Taiwan shifted production to concentrate more and more of their own investment (and their government's attentions) on industrial electronics, as Chin Chung shows in the next chapter.[24] As these developments occurred, there appeared an ever more elaborate and deepening technical division of labor between U.S.-based and Asian-based operations, bound together in production networks serving U.S. firms' advanced-country markets. In essence, a new supply base was being created in Asia under the control of U.S. and local, but not Japanese, capital.

Indeed, while Asia's indigenous electronics capabilities (excluding Japan) developed in close symbiosis with the strategies and activities of U.S. multinational firms, they were driven by local private investment and supported by government policies. Outside South Korea (where the *chaebol* dominated domestic electronics development), resident ethnic Chinese investors played the principal private entrepreneurial role in the China Circle, Singapore, and later in Malaysia, Indonesia, and Thailand. During this period, in the newly industrializing economies, or NIEs (and later in Southeast Asia) governments provided a panoply of fiscal and tax incentives, invested heavily in modern infrastructure, generic technology development, and improvement of the technical skills of the work force, engaged in selective strategic trade interventions, and sometimes even provided market intelligence and product development road maps.[25] The aims were both to plug into the developing multinational production networks in the region and to use them as a lever toward autonomous capabilities. The result, by the end of the 1980s, was burgeoning indigenous electronics production throughout the region, with most of it, outside Korea, under the control of overseas-Chinese (OC) capital.[26]

Not surprisingly, given their deep ties in this period to U.S. producers, OC firms were concentrated in the PC and PC-related product markets. In turn, the nerve centers of OC activity in PC electronics were Taiwan and Singapore, the home bases for emerging Asia-Pacific multinationals like Acer (Taiwan) and Creative Technologies (Singapore).[27] Taiwanese producers were at the heart of the nascent alternative supply base. Ultimately, their position would crystallize in the third period, culminating by the mid-1990s, as table 5-1 shows, with significant-to-dominant world market shares in fourteen PC-related supply categories. Singapore-based OC producers similarly

Table 5-1. *World Market Share of Taiwanese Firms in PC-Related Products, 1994*[a]

Product	*Share*
Motherboard	80
Mouse	80
Scanner	61
Monitor	56
Keyboard	52
Network interface card	34
Graphics card	32
Switching power supply	31
Notebook PC	28
Video card	24
Terminal	22
Network hub	18
Audio card	11
Desktop PC	8

Source: Drawn from a presentation prepared by Tze-Chen Tu, director of Taiwan's Market Intelligence Center of the Institute for Information Industries, "Upgrading Taiwan's IT Industry—New Challenges and the Role of International Cooperation," at the BRIE–Asia Foundation Conference on Competing Production Networks in Asia Host Country Perspectives, San Francisco, April 27–28, 1995.

began to emerge in this second period as significant suppliers of hard disk drive–related components and services and of multimedia sound cards, PC subassemblies, and PC assembly services.[28]

By contrast to both the U.S. and OC developments in this second phase, the pattern of Japanese investment led to a dual production structure controlled by Japanese firms and premised on traditional product cycles. Sophisticated products were produced at home with sophisticated processes to serve advanced-country markets, while lower-end products were produced with simple processes in regional affiliates to serve local Asian markets. Both sets of operations sourced from a common supply base, located largely in Japan and controlled, directly or indirectly, by Japan's major electronics companies. Where Japanese companies responded to government or commercial pressures to localize, they did so, as Ernst suggests, from within their established supply base—that is, by transplanting the operation of an affiliated domestic Japanese supplier—not by sourcing locally from the emerging Asian supply base. In short, the Japanese production networks boasted redundant investment and remained relatively closed, even as the U.S. networks became more open, more entwined with indigenous OC producers, and more specialized.

The Third Stage

These trends were fully elaborated during the third stage, from 1985 to the early 1990s. At home, U.S. firms focused scarce corporate resources more intensely on new product definition and the associated skills (such as design, architectures, and software) necessary to create, maintain, and evolve de facto market standards. In turn, they upgraded their Asian affiliates, giving them greater responsibility for hardware value-added and manufacturing, and significantly increased local sourcing of components, parts, and subassemblies. They even contracted out the design and manufacture of some boards and components. Thus during this period the Asian affiliates of U.S. firms continued to migrate from printed circuit boards (PCBs) to final assembly with increased automation; to increase both component production and final system value-added; and to assume global responsibility for higher value-added systems (for example, from monochrome desktops to color notebook PCs). Their production networks extended to more and more capable local Asian producers, which became increasingly skilled suppliers of components, subassemblies, and sometimes entire systems. Even in areas like memory chips and displays, in which Japanese firms remained important suppliers to U.S. firms, there was sufficient competition from other Asian sources (such as Korea, in memory chips) or sufficient political pressure to keep the supply architecture open.

Major U.S. producers of PCs like Apple illustrate well these developments.[29] Apple Computer Singapore (ACS) opened a PCB assembly plant for the Apple II PC in 1981. By 1983 nine local companies were contract manufacturing PCBs for the Apple IIe and Lisa PCs. By 1985 ACS was upgraded to include final assembly of Apple IIes for the world market. From 1986 to 1989 ACS was expanded and upgraded to begin some component design work. In 1990 ACS assumed final responsibility for assembling two of three new Macintosh PCs (and PCBs for the third) and designed (locally) and manufactured associated monitors. By then, essentially all components were sourced in Asia (except the U.S.-made microprocessor); ACS's 130 main suppliers included local firms like Gul Technologies and Tri-M (PCBs). ACS had also demonstrated that its growing technical prowess could pay competitive dividends in speeding time to market. It was able to move from design to production rollout in up to half the time of Apple's other facilities. By 1992 ACS assumed responsibility for final assembly for all Asia-Pacific markets, including Japan, was designing and supplying boards globally, manufacturing monitors and some peripherals, and designing chips. More than $1 billion* was being procured annually through ACS. In 1993 ACS set up a design center for high-volume desktop products, Apple's only hardware design center outside the United States. By 1994 ACS had become the center for distribution, logistics, sales, and marketing for the Asia-Pacific region and was assembling the Mac Classic II, LC III and IV, midrange Centris, and Quadra 800 for global distribution. Regional sourcing reached $2 billion, half from Japan (liquid crystal displays [LCDs], peripherals, memory, hard disk drives), one-quarter from Singapore, and $250–$500 million from Taiwan for OEM desktops, monitors, PCBs, powerbooks, digital assistants, and chips. Korea's Goldstar also supplied monitors. By late 1994 ACS had begun to design the motherboard and tooling for the multimedia system Mac LC 630 PC and to assemble it for worldwide export. Two new Mac products completely designed and manufactured at ACS were launched in 1995.

The value-added and local-sourcing progression of other major U.S. electronics players in Asia is broadly similar.[30] For example, Compaq Asia (hereafter CAS, for Compaq Asia-Singapore) established its Singapore factory in 1986 for PCB assembly of components sourced from Asia (including Japan) and for desktop PCs to be final assembled in the United States. By 1994, after terminating an OEM relationship with Japan's Citizen Watch, CAS was designing and manufacturing all notebook and portable PCs for worldwide consumption and all desktop PCs for the Asia-Pacific region. Similarly, Hewlett-Packard's Singapore operations evolved from the assembly of calculators in 1977 to global responsibility for portable printers and Pentium desktop PCs and servers, with local manufacturing, process design, tooling development, and chip design. Motorola's Singapore operations evolved from simple PCB assembly of pagers and private radio systems destined for the United States in 1983, to worldwide mandates for design, development, and automated manufacture of double-sided six-layer PCBs, for design and development of integrated circuits for disk drives and other periph-

*All dollar amounts are U.S. dollars unless otherwise indicated.

Table 5-2. *Taiwan Firms' Original Equipment Manufacturing (OEM) Relations in PC-Related Products, 1994*[a]

OEM producer	*Buyer*	*Product*
Acer	Apple, Fujitsu, NEC,[b] NCR, Data General, Siemens	Notebooks and/or monitors
Delta	Apple, Compaq, IBM	Power supplies
Elite	DEC,[c] IBM, NEC,[b] Siemens	Motherboards
FIC[d]	AT&T, Dell, Unisys	Motherboards
Inventa	Apple, Compaq, Dell	PDA, notebooks
Lite-on	Compaq, DEC,[c] Dell	Power supplies and/or monitors
Tatung	Apple, Packard Bell, NEC[b]	PCs and/or motherboards, monitors

Source: Tu, "Upgrading Taiwan's IT Industry," and press reports.
a. A representative sample.
b. Netscape Communications.
c. Digital Equipment Corporation.
d. First International Computer.

erals, for some research and development, and for sourcing of at least $500 million of parts and components within the region. Similar kinds of stories could be told for AT&T in telecommunications products; for IBM and Digital Equipment in PCs and peripherals; for Maxtor, Connor, Seagate, and Western Digital in hard disk drives; and for Texas Instruments, Intel, and National Semiconductor in semiconductors.

As U.S. Asia-based affiliates upgraded and specialized in this way during the third period, their indigenous OC suppliers followed suit. Table 5-2 gives some indication of this by examining the emergence, by the period's end, of OEM relationships with major China Circle producers. In turn, by leveraging their link into the U.S. production networks and the global distribution capabilities thereby provided, the strongest China Circle producers began to control their own production networks. In her chapter Chung shows how, in the early 1990s, intense competition and growing needs for scale-intensive investment forced a shakeout and consolidation among Taiwanese- and Hong Kong-based electronics firms.[31] Firms like Acer, the Formosa Plastics Group, and Tatung began to ride herd on an extensive indigenous supply base of thousands of small and medium-size design, component, parts, subassembly, and assembly houses throughout the China Circle and extending into Southeast Asia. These firms form an intricate subcontracting structure of affiliated and family enterprises that constitute the

Table 5-3. *Domestic and Offshore Production Value of Taiwan's Electronics Industry, 1992–95*
Millions of U.S. dollars

Production	*1992*	*1993*	*1994*	*1995 (est.)*
Domestic	8,391	9,693	11,579	13,139
Offshore	973	1,691	3,003	4,279
Offshore/domestic (percent)	11.6	17.5	25.9	32.6

Source: Tu, "Upgrading Taiwan's IT Industry."

local production network and supply base. The many small firms are aligned vertically with the few larger-scale enterprises that act as intermediaries for foreign MNC customers.[32] Designs and key components flow down from the large-scale enterprises; more labor-intensive production activities flow up along the subcontract network leading to final assembly.

Toward the end of the third period, because of steep rises in factor input costs in the NIEs and by currency appreciation, these emerging OC production networks become more and more regionalized. For example, table 5-3 suggests the extent to which considerable PC-related production is now being carried on by Taiwanese MNCs within the region but outside Taiwan. As the table shows, production outside Taiwan but in the OC Asian networks accounted for increasing shares of total production under Taiwanese control, approaching one-quarter of the total in 1995. As Chung argues, the offshore activity is concentrated in certain product segments, with about two-thirds of "Taiwanese" production of keyboards, one-half of power supplies, and about one-quarter of monitors and motherboards now taking place outside Taiwan. Investment targeted both mainland China and Southeast Asia—partly a result, as other chapters in this volume suggest, of the timing of both Taiwanese and mainland policy reforms, and partly of prudent geographic risk-spreading by OC investors.[33]

In sum, by the early 1990s the division of labor between the United States and Asia, and within Asia between affiliates and local producers, deepened significantly, and U.S. firms effectively exploited increased technical specialization in Asia. In stark contrast, up through the end of 1993 Japanese firms still controlled their Asian affiliates' major decisionmaking and sourcing activities from Japan. More low-end process and product technology had been moved offshore, including production of audio systems (cassette recorders, headphones, low-end tuners, and so on), under-20-inch televisions, and some VCR models, cameras, calculators, and appliances such as microwave ovens. Local Asian content had risen toward 60 percent, but core technological inputs like magnetrons, chips, and recording heads were exclusively sourced from Japan, and the 60 percent "local" content was mostly supplied by the offshore branch plants of traditional domestic Japanese suppliers. Local design activities invariably served to tailor Japanese product concepts for local Asian markets; global mandates for advanced products, let alone their design, development, and manufacture, were nowhere to be found outside Japan. In contrast to U.S. producers, for example, Japanese PC producers sourced displays, memory, some microprocessors, drives, power and mechanical components, plastics, and PCBs from Japan (or in the case of some low-end components, from offshore affiliates) and did PCB and final assembly and essentially all advanced design and development in Japan. In short, Japanese firms intensified rather than rationalized their dual production structure, and by excluding others from their production networks, failed to benefit from the increasing, cheaper, and faster technical capabilities in the rest of Asia.

A Network Typology and the Future of Competition

In Asia today, beneath the superficial similarity engendered by aggregate trade and investment data and macroanalyses, lie distinctly different electronics production networks under the control of U.S., Japanese, and OC multinationals. The U.S. networks tend to be open to outsiders, fast and flexible in decisionmaking and implementation, structured through formal, legal relationships, and capable of changing contour (and partners) as needs change—in essence, open, fast, flexible, formal, and disposable. Their activities are centered in the NIEs, especially Singapore, but increasingly reach into the rest of Asia and China. By contrast, the Japanese networks tend to be relatively closed to outsiders, more cautious to make and implement decisions, which are generated from Japan, and structured on stable, long-term business, and *keiretsu* relationships—that is, closed, cautious, centralized, long term, and stable. Despite the recent surge of Japanese investment in Asia, their networks are still centered in Japan.

The respective networks also rely on distinctively different supply bases, boast different product mixes, and, most important, constitute very different divisions of labor. The U.S. networks rely on an open, competitive supply architecture in which Japanese, U.S., Taiwanese, Singapore, Korean, and other Asian firms compete on cost, quality, and time to market and sometimes provide significant value-added. By contrast, the Japanese networks rely on a largely domestic and affiliated supply base with little value-added by other Asian producers. The U.S. networks produce (and in some cases design and develop) increasingly sophisticated industrial electronics like hard disk drives, PCs, InkJet printers, and telecommunications products. The Japanese networks still mostly produce consumer audio-visual electronics and appliances. The U.S. networks exploit a complementary division of labor in which U.S. firms specialize in "soft" competencies (definition, architecture, design-standards areas) and Asian firms specialize in hard competencies (components, manufacturing stages, and design and development thereof). By contrast, the Japanese networks exploit a division of labor with significant redundancies in which domestic Japanese operations produce high-value, high-end products using sophisticated processes, and offshore affiliations produce low-value, low-end products. The U.S. networks exploit increasing technical specialization throughout the production process in which the Asian contribution is maximized; the Japanese networks exploit a value-added specialization between products in which the Asian contribution is minimized.

By comparison, the emerging OC networks appear to combine features of both the Japanese and U.S. MNC approaches, with distinctive characteristics of their own. Much like the Japanese ones, OC networks are difficult for outsiders to penetrate. Much like the U.S. ones, OC networks are fast and flexible. Indeed, industry estimates of OC network business speed peg the time from conception to execution at a fraction of that of larger MNCs burdened with formal organization and layered decisionmaking.[34] In some instances, OC networks can design and execute in less time than it takes the Japanese giants just to make a go-ahead decision.[35] For the Taiwanese design houses, in particular, this capability is apparently built on a high-value-added foundation, macrocell-based design methodologies, and libraries of already characterized component functions that can be combined and altered to implement new concepts.[36] The rapid design capability then joins with the hypercompetition among subcontractors in the network to implement the new designs as fast as possible.

Unlike either the U.S. or Japanese networks, the OC networks seem especially focused on the intricate division of production tasks (such as components and subassembly steps) that can be farmed out, down to family job shops and home workers. Individual units within the network operate at small scale with minimal capital investment requirements and link on the informal bases of *guanxi*. The flexibility that results, mirroring the industrial district capabilities in Italy and parts of Germany, makes it possible to increase or decrease production scale on short notice or to enter and exit niche-product-market segments, all at minimal cost and with minimal fixed investments.[37]

The best OC networks also run extremely lean in general, sales, and administrative overheads, in which they match the best practices of MNC leaders like Hewlett-Packard (at about 10 percent of sales for microcomputers and printers) and are far superior to most advanced MNC performers (15 percent to upward of 20 percent of sales).[38] Of course, such cost minimization is inherent in the subcontract structure of the OC production networks, in which affiliates and family enterprises can be squeezed (if necessary, in a time-honored sweatshop manner).

In short, the OC networks appear to be insular, fast, flexible, *guanxi*-mediated, and fluid. They tend to be centered in the China Circle and increasingly focused on mainland China. Like their U.S. counterparts, the OC networks seek to exploit a highly competitive supply base and concentrate on industrial electronics. Much like the Japanese, OC networks retain in the home base high-value-added products manufactured with more advanced processes and move offshore to cheaper production locations lower-value-added products assembled with simpler processes. Unlike the Japanese, however, the OC networks also self-consciously leverage increasing technical specialization through local relationships wherever possible. And unlike the relationships in the other two networks, the OC network relationships are increasingly China-centered. Rather than use a NIE base as the regional center, OC networks may end up with a China base as their global center, using demand and technical know-how in the domestic China market to achieve world-class scale, costs, and innovation.

As argued at the outset, the competitive consequences of the differences among U.S., Japanese, and OC networks have been significant. The U.S. networks relieved the constraining threat of competitive dependence on their Japanese rivals by reconstituting the architecture of supply in electronics. Simultaneously, the turn to skilled but cheaper

Asian suppliers helped to lower total production costs, fierce competition within the supply base helped to reduce turnaround times, and specialization and diversity within the network permitted U.S. producers to keep better pace than their Japanese rivals with rapid technological and market shifts. Growing Asian technical capabilities freed U.S. firms to focus their efforts (and scarce resources) on new product definition and standards competitions, systems integration, software value-added, and distribution. In the bargain, the U.S. networks helped to spawn and sustain direct Asian competition to Japanese firms in several of their stronghold markets, such as memory chips, consumer electronics, and LCDs. And while OC network capabilities grew prodigiously, they did not directly challenge revived U.S. leadership in the last round of competition. Overall, U.S. firms not only stayed the competitive course but also prospered.

Yet the current U.S. position is no more a guarantee of future success than was Japan's in the early 1980s. Much depends on how Japanese firms respond to their present competitive dilemmas and on how OC firms leverage opportunities in the China Circle.

The Ernst and Huchet chapters provide evidence of nascent Japanese adjustment that, at least at first blush, appears to draw a different image from the closed network structure emphasized here. The authors see some evidence of increased openness and increased reliance on OC, Chinese, and Korean suppliers as Japanese firms adjust to the competitive success of U.S. and indigenous Asian producers and target the China market. Whether those changing characteristics are permanent or temporary is an open question, however. In 1996 there was anecdotal evidence that Japanese networks were snapping back toward the more traditional, closed model as the yen again depreciates and as Japanese firms absorb know-how from the partners they took on in Asia. In any case, there is no evidence that the basic Japanese strategy of controlling value-added through ownership has changed; nor do Japanese firms appear intent on exploiting increased specialization in the rest of Asia wherever they can do the specialization themselves.[39]

Ernst and Huchet would also agree that the precise characteristics of Japan's Asia-based networks that created vulnerability over the last decade—closed, cautious, Japan-centered, long-term and stable—could be turned into competitive strengths with a dose of rationalization and a pinch of vision. Japanese firms could decide to accept slower domestic growth and the need to exploit technical capabilities in the rest of Asia as givens. They could decide to selectively incorporate Asian producers into the family and build stable, long-term, mutually advantageous ties focused on exploiting specific technological capabilities in other parts of Asia. They could decide to invest for the long term. They could decide to drive their growth from Asia's growth. If Asia becomes a launch market for new product concepts—and its rapid growth and burgeoning wealth suggest that it must in some market segments—Japanese firms might then be better positioned to exploit the development.[40]

Just as big a competitive wild card is the growing electronics capability in the China Circle linked to OC investments in the United States, Southeast Asia, and eventually Japan. A competitive China Circle scenario is easy enough to describe. The combination of Hong Kong-based financial and producer services with Taiwan-based digital product and process design, Southeast Asian component specialization, and highly skilled but cheap mainland labor—and of course the mainland market—provides a tantalizing scenario for regional dominance. The OC network characteristics identified above—insulated from outside control, fast, flexible and fluid—appear to be a compelling mix for exploiting the region's possibilities. And the sheer scale of production for the mainland and, from the mainland, for overseas markets would dwarf the leverage provided by any other home market base. To this potent brew should be added the self-conscious developmental intent of governments throughout the region to nurture indigenous capabilities, and of China's to move to the technological frontier as fast as possible.

The significant constraints on the emergence of such a scenario should not be underestimated, however. Unlike the U.S. networks, which have retained capability in most core component technologies and an important though diminished position in capital goods, the OC networks remain dependent on Japanese competitors for advanced manufacturing equipment and high-value-added core components (for example, for Taiwanese producers, $500 million of LCDs and $3 billion of memory chips in 1994). Even more of a constraint, however, is continuing dependence on the American networks for microprocessor architectures, advanced product concepts, and global distribution. It is likely that the Chinese market can eventually help to break those constraints—by providing the returns to invest to relieve core component dependence, the new product concepts that can become global standards, and leverage to develop indigenous brands and global distribution channels. But that is likely to take time, probably several decades.

In the interim the China Circle will witness one of the great market battles in memory as U.S., Japanese, and indigenous production networks vie for twenty-first century advantage.

Notes

1. The government's support took two forms—direct financial support of $100 million a year to the industry's manufacturing technology consortium, Sematech—or half of Sematech's annual budget—and negotiation of the U.S.-Japan Semiconductor Trade Agreement. For details, see Michael Borrus, *Competing for Control: America's Stake in Microelectronics* (Ballinger, 1988). For an elaborate analysis of the agreement, see Kenneth Flamm, *Mismanaged Trade? Strategic Policy and the Semiconductor Industry* (Brookings, 1996).

2. This position is argued explicitly by industry consultant William F. Finan and his academic collaborator Jeffrey Frey in *Nihon no gijyutsu ga abunai: kenshō, haiteku sangyō no suitai* (Japan's crisis in electronics: failure of the vision) (Tokyo: Nikkei Press, 1994).

3. Steven Jay Gould makes this point with respect to natural selection in *Wonderful Life* (Norton, 1989), p. 236.

4. For a longer analysis, see Borrus, *Competing for Control*. The domestic market served as a launch market during the late 1970s and 1980s for, among other products, the VCR, Camcorder, Walkman, hand-held TV, fax machine, portable copier, and notebook PC.

5. On the latter point, see Ichiro Uchida, "Restructuring of the Japanese Economy," in Eileen Doherty, ed., *Japanese Investment in Asia: International Production Strategies in a Rapidly Changing World* (University of California, Berkeley: Berkeley Roundtable on the International Economy, 1995). (Hereafter BRIE.)

6. On the former point, see Uchida, "Restructuring of the Japanese Economy." As argued later—and somewhat at odds with Ernst's perspective—rationalization does not necessarily imply radical change in the way Japan's industrial firms operate. For example, some Japan scholars like Greg Noble argue that Japan's lifetime employment system appears to have survived the recent economic shocks essentially intact, with only marginal modifications. Verbal remarks of Noble at the BRIE-Asia Foundation Conference on Competing Production Networks in Asia: Host Country Perspectives, San Francisco, April 27–28, 1995.

7. By "open" I mean that key product specifications, especially the interface specifications that permit interoperability with the operating system or system hardware, are published or licensed and thus available to independent designers of systems or software who can produce complementary or competing products.

8. By "production network" I mean the organization, across national borders, of the relationships (intrafirm and increasingly interfirm) through which the firm accomplishes the entire value chain of production, including research and development, product definition and design, supply of inputs, manufacturing, distribution, and support services. Especially significant in Asia are supplier relationships that include subcon-

tracting, OEM (original equipment manufacturing), and ODM (original design manufacturing) arrangements between foreign MNCs and domestic suppliers of intermediate production inputs, such as materials, tools and molds, parts and components, subassemblies, and software—some of whom may also compete in final product markets. See the elaboration in Dieter Ernst, "Networks, Market Structure, and Technology Diffusion: A Conceptual Framework and Some Empirical Evidence," report prepared for the Organization for Economic Cooperation and Development, Paris, 1992.

9. The next section defines the concept of supply base.

10. By "continuous innovation" I mean the capacity to add incremental advances in performance, functionality, or features within or between given product generations—for example, from 75 MHz to 250 MHz Pentium microprocessors, or from 25 MHz 386-based PCs with 4 megabytes of random-access memory (RAM) and 100 megabyte hard drives to 200 MHz Pentium-based PCs with 32 megabytes of RAM, 2 gigabyte hard drives, and CD-ROM drives.

11. Applying and exploring the limits of this method are principal goals of BRIE research, supported by the Alfred P. Sloan Foundation.

12. See the discussion in Laura Tyson, "They Are Not Us," *American Prospect,* no. 4 (Winter 1991), pp. 37–49. See also Yao-Su Hu, "Global Corporations Are National Firms with International Operations," *California Management Review,* vol. 34 (Winter 1992), pp. 107–26. Of course, the debate addressed by these articles was popularly launched by Robert Reich in "Who Is Us?" *Harvard Business Review,* vol. 90 (January–February, 1990), pp. 53–64. More generally, the recent volume edited by Dennis Encarnation and Mark Mason supports the persistence of important differences among multinational firms based on national origin; see Encarnation and Mason, eds., *Does Ownership Matter?* (Oxford University Press, 1995).

13. See "A Survey of Multinationals," *Economist,* March 27, 1993, pp. 6–7, citing United Nations data. The major exceptions are oil companies (because oil fields tend to be located abroad) and small-country multinationals like Nestle, Unilever, and ABB (because their markets are located abroad)—and the latter would fall into the 60 to 90 percent range if Europe were treated as their home base. By that measure, the most international non-oil MNC is IBM, with about 50 percent of assets outside the United States. But because half of its assets are still concentrated in the United States, even IBM can be said to have the United States as its home base.

14. This conclusion is easily reached from industry conversations and even a quick perusal of the annual reports of the 1,000 largest U.S. and 1,000 largest non-U.S. firms. More generally, the evidence in John Dunning's comprehensive work on MNCs supports this conclusion, as does Michael Porter's work. See John Dunning, *Multinationals, Technology, and Competitiveness* (London: Unwin Hyman, 1988); and Michael Porter, *The Competitive Advantage of Nations* (London: Macmillan, 1990).

15. My colleague Stephen Cohen coined the concept of national market logics in Michael Borrus and others, "Globalization and Production," Working Paper 45 (BRIE, 1991).

16. For a discussion of this concept of technology trajectories, see Michael Borrus, "The Regional Architecture of Global Electronics: Trajectories, Linkages, and Access to Technology," in Peter Gourevitch and Paolo Guierrieri, eds., *New Challenges to*

International Cooperation: Adjustment of Firms, Policies, and Organizations to Global Competition (San Diego: University of California, 1993).

17. For one effort to elucidate some of these variables—the state, labor relations, and financial systems—as part of a formal analysis explaining national economic development, see John Zysman, "How Institutions Create Historically Rooted Trajectories of Growth," *Industrial and Corporate Change*, vol. 3, no. 1 (1994), pp. 243–83. Some of the variables described earlier are similar to those used by Porter, *Competitive Advantage,* but to different ends in a decidedly different, albeit complementary, analysis.

18. On the *guanxi* network concept, see Gary G. Hamilton, "Competition and Organization: A Reexamination of Chinese Business Practices," paper prepared for the IGCC Conference on the China Circle: Regional Consequences of Evolving Relations among the PRC, Taiwan, and Hong Kong-Macao, Hong Kong, December 8–10, 1994.

19. For a more detailed discussion of this U.S.-Japan comparison, see Borrus, *Competing for Control,* chaps. 4, 5.

20. See the discussion of sequential increasing supply dependence in consumer electronics in Consumer Electronics Sector Working Group, "The Decline of U.S. Consumer Electronics Manufacturing: History, Hypotheses, Remedies," MIT Commission on Industrial Productivity, Cambridge, Mass., December 1988.

21. For an extended discussion of the supply base and architecture of supply concepts, see Borrus, "Regional Architecture of Global Electronics."

22. For the broad range of major component technologies involved, see the discussion in Borrus, "Regional Architecture of Global Electronics."

23. The characterization of U.S. foreign direct investment is based on the BRIE U.S. electronics FDI database, compiled from public sources and maintained by Greg Linden, supplemented by industry conversations, and reviewed by senior managers with Asia responsibility from most of the firms mentioned in the text. The characterization of Japanese electronics FDI in Asia that follows is consistent with, and in part draws on, data and details in the chapters by Ernst and Huchet, as well as from Ken-ichi Takayasu and Yukiko Ishizaki, "The Changing International Division of Labor of Japanese Electronics Industry in Asia and Its Impact on the Japanese Economy," *RIM: Pacific Business and Industries*, vol. 1, no. 27 (1995), pp. 2–21. See also the relevant chapters in Doherty, *Japanese Investment in Asia.*

24. On the progression from consumer to industrial electronics in Taiwan, see also Scott Callon, "Different Paths: The Rise of Taiwan and Singapore in the Global Personal Computer Industry," Japan Development Bank Discussion Paper Series 9494 (Tokyo, August 1994). More generally, on the development of Taiwan's information technology industry, see Kenneth L. Kraemer and Jason Dedrick, *Entrepreneurship, Flexibility, and Policy Coordination: Taiwan's Information Technology Industry* (Irvine, Calif.: Center for Research on Information Technology and Organizations, 1995).

25. There were of course tremendous variations in the role played by state policy, and in the policies themselves, in the different countries of the region. In highlighting a few commonalties, I do not mean to slight those differences. The active role played in general by governments in the region has been explored in detail in many scholarly works. See, for example, Robert Wade, *Governing the Market: Economic Theory and the Role of Government in East Asian Industrialization* (Princeton University Press,

1990); and Stephan Haggard, *Pathways from the Periphery: The Politics of Growth in the Newly Industrializing Countries* (Cornell University Press, 1988). More recently, see the excellent contributions in Andrew MacIntyre, ed., *Business and Government in Industrializing Asia* (Cornell University Press, 1994).

26. In focusing on OC electronics producers, I am ignoring the significant regional investment by the Korean *chaebol*, which emerged during this period as major, region-wide producers of consumer electronics and components. See, for example, Martin D. Bloom, "Globalization and the Korean Electronics Industry," *Pacific Review*, vol. 6, no. 2 (1993), pp. 119–26.

27. This picture is perhaps a bit at odds with the emphasis in the rest of this volume on the importance of Hong Kong as a force for China Circle integration. Hong Kong was at best a distant third most important NIE site for MNC or indigenous electronics production. Moreover, its relative importance declined as investment spread into Southeast Asia and mainland China—though Hong Kong undoubtedly played an important role in helping to channel Taiwanese investment onto the mainland during the third period, described later.

28. Callon, "Different Paths."

29. Based on press accounts, company annual reports, and SEC 10K filings, as compiled by Greg Linden. See Linden, "Apple Computer East Asian Manufacturing Affiliates," November 7, 1994, BRIE Asia FDI database.

30. For Compaq, see Linden, "Compaq East Asian Manufacturing Affiliates," November 7, 1994; for Hewlett-Packard, see Linden, "Hewlett-Packard East Asian Manufacturing Affiliates," November 9, 1994; for Motorola, see Linden, "Motorola East Asian Manufacturing Affiliates," November 7, 1994—all from BRIE Asia FDI database.

31. As Chung shows, the resulting industry concentration was most visible in Taiwan's largest domestic product sectors, notably monitors, PCs, and PCBs, where the top ten indigenous producers now account for over 70 percent of the market.

32. For elaboration on the following, see, for example, G. S. Shieh, "Network Labor Process: The Subcontracting Networks in Manufacturing Industries of Taiwan," *Academia Sinica* (Bulletin of the Institute of Ethnology), no. 71 (Spring 1991); Brian Levy and Wen-Jeng Kuo, "The Strategic Orientation of Firms and the Performance of Korea and Taiwan in Frontier Industries: Lessons from Comparative Case Studies of Keyboard and Personal Computer Assembly," *World Development*, vol. 19, no. 4 (1990), pp. 363–74. I have also drawn on an excellent paper by one of my graduate students, Fu-mei Chen, "From Comparative Advantage to Competitive Advantage: A Case Study of Taiwan's Electronics Industry," May 21, 1994, University of California, Berkeley.

33. The Taiwan Strait crisis of 1996 does not appear to have substantially influenced the willingness of Taiwanese electronics firms to invest on the mainland, though it undoubtedly reinforcd their desire to spread the risk by simultaneously investing in Southeast Asia.

34. Representative estimates range from Apple's judgment that its Singapore operation can move a new product into production in half the time of its other operations, to Ming Chien, chairman of First International Computer, who estimates that motherboards can be completely changed out (with all attendant alterations to the rest of the system) in Taiwan in two to three weeks, as against up to a year in the United States.

On Apple's judgment, see Singapore *Business Times*, November 27, 1990, p. 14; on Ming Chien, see Callon, "Different Paths."

35. Louis Kraar, "Your Next PC Could Be Made in Taiwan," *Fortune*, August 8, 1994, pp. 90–96, inferring from Dataquest estimates.

36. See Callon, "Different Paths," citing interviews in Singapore. Structured integrated circuit (IC) design approaches were pioneered in the United States at universities like Berkeley and CalTech, where many OC engineers were formally trained.

37. On European industrial districts, see the work on flexible specialization; notably, Michael Piore and Charles Sable, *The Second Industrial Divide: Possibilities for Prosperity* (Basic Books, 1984).

38. Based on industry discussions.

39. So long as Japanese, U.S., and OC firms continue to be driven by very different domestic linkages, strategies, industrial structures, and policy, capital market, and labor market influences, their network differences are likely to persist even if they converge in competitive purpose.

40. In fact, the opportunity to drive development out of Asia is already appearing in a set of significant potential product markets. These include broadcast media, in which firms like Hong Kong's TVB and Murdoch's Star TV are pioneering direct broadcast TV transmission; software, in which indigenous concepts could lead in new directions; and segments of the wireless communication markets, in which, for example, Motorola projects that China will pass the United States to become its largest market for pagers in the next few years.

CHAPTER SIX

Division of Labor across the Taiwan Strait: Macro Overview and Analysis of the Electronics Industry

Chin Chung

BOTH TAIWAN AND CHINA underwent significant structural changes during the 1980s. Taiwan, as one of the leading newly industrializing economies (NIEs) in the world, faced growing labor shortages and a rapidly appreciating currency in the mid-1980s as part of the unpleasant costs of accelerated economic growth. For Taiwan to begin a new phase of development, it needed to upgrade its technology and reorient its industries. China, by contrast, awakened from decades of Maoist communism and embarked on massive economic reforms in the late 1970s. The core of these reforms lies in two strategies: "opening up to the world" and "fostering internal economic dynamism." In following these principles, China has implicitly adopted an export-oriented policy based on foreign direct investment (FDI) combined with indigenous labor resources to achieve high economic growth—a strategy similar to one that has made the Four Little Dragons (Hong Kong, Singapore, South Korea, and Taiwan) prosperous.

In this chapter I analyze the emerging pattern of the division of labor between China and Taiwan, using systematic data on trade, investment, and production to envisage a dual relationship between the two economies: as competitors in the world marketplace and cooperative partners in a wide array of manufacturing activities. In particular, I closely examine recent developments in one of the key sectors in Taiwan's economy, the electronics industry, to gain a clearer picture of the new forms of economic interaction across the Strait. These developments not only may change the bilateral economic relationship between Taiwan and China but may also strongly influence the reshaping of the broader economic landscape beyond the region.

The Emerging Pattern of Division of Labor

During the three decades before 1986, Taiwan successfully built up its industrial capability primarily through an export-led growth policy. In 1987 the Taiwanese government abolished martial law and lifted the ban on kinship visits to the mainland after a forty-year freeze, while also liberalizing the foreign exchange regime. The impact of these policy changes on the course of Taiwanese FDI toward China can be easily discerned from table 6-1. Compared with ASEAN countries, China has been more successful in attracting FDI over the past decade. From 1986 to 1994 China accumulated some $288 billion* in FDI inflow, more than four times the amount Indonesia, the second largest FDI recipient in the region, received. Taiwanese FDI on the mainland quickly took off after 1987 and soon overshadowed its FDI in ASEAN countries. Moreover, since 1991 FDI from Japan and the United States has gradually shifted from ASEAN countries to the PRC.

Being politically and culturally intertwined with Hong Kong and Taiwan, China has naturally drawn most of its inward FDI from its two cultural siblings. But other countries, including the United States and Japan, also have shown great interest in the Chinese market and have invested quite heavily in China, albeit still with a sense of caution. The motivations for investing in mainland China are likely to be different, though, for the advanced countries than for a NIE like Taiwan. China, as a host to FDI, not only possesses the obvious location advantage of low-cost labor and land but also has the additional advantage of a vast and rapidly growing internal market.[1] Though the latter attribute is enticing for both advanced countries and NIEs, the first advantage is probably more important to NIEs seeking to relocate their traditional

---

*All dollar amounts are U.S. dollars unless otherwise indicated.

Table 6-1. *Cumulative Foreign Direct Investment in China and Three ASEAN Countries, 1986–94*

Millions of U.S. dollars; numbers in parentheses are percent shares

Recipient and date	*Investor*				
	World total	*Hong Kong*	*Taiwan*	*Japan*	*United States*
China					
1986	2,834 (100)	1,449 (51.1)	0 (0.0)	210 (7.4)	530 (18.6)
1987	3,709 (100)	1,947 (40.4)	100 (2.7)	301 (8.1)	342 (9.2)
1988	5,297 (100)	3,467 (65.5)	421 (7.9)	276 (5.2)	370 (7.0)
1989	5,599 (100)	3,160 (56.4)	523 (9.3)	439 (7.8)	641 (11.4)
1990	6,597 (100)	3,676 (55.7)	984 (14.9)	458 (6.9)	357 (5.4)
1991	11,977 (100)	7,215 (60.2)	1,358 (11.3)	812 (6.8)	548 (4.8)
1992	58,123 (100)	40,044 (68.9)	5,543 (9.5)	2,173 (3.7)	3,121 (5.4)
1993	111,710 (100)	78,650 (70.4)	9,450 (8.5)	1,450 (1.3)	6,300 (5.6)
1994	82,680 (100)	46,971 (56.8)	5,395 (6.5)	4,440 (5.4)	6,010 (7.3)
1986–94	288,252 (100)	181,868 (63.1)	24,289 (8.4)	12,069 (4.2)	18,729 (6.5)
Indonesia					
1986	n.a.	n.a.	18 (n.a.)	n.a.	n.a.
1987	1,520 (100)	129 (8.1)	8 (0.6)	554 (35.4)	80 (5.3)
1988	4,482 (100)	253 (5.6)	914 (20.4)	391 (8.7)	671 (15.0)
1989	4,719 (100)	407 (8.6)	157 (3.3)	779 (16.5)	348 (7.4)
1990	8,751 (100)	993 (11.4)	618 (7.1)	2,241 (25.6)	154 (1.8)
1991	8,778 (100)	278 (3.2)	1,056 (12.0)	929 (10.6)	276 (3.1)
1992	10,313 (100)	1,018 (9.9)	563 (5.5)	1,502 (14.6)	922 (8.9)
1993	8,144 (100)	384 (4.7)	131 (1.6)	836 (10.3)	444 (5.5)
1994[a]	23,092 (100)	6,013 (26.1)	2,480 (10.7)	1,388 (6.0)	445 (1.9)
1987–94	69,889 (100)	9,475 (13.6)	4,927 (7.0)	8,620 (12.3)	3,778 (5.4)
Malaysia					
1986	654 (100)	22 (3.3)	4 (0.6)	45 (6.9)	21 (3.2)
1987	818 (100)	35 (4.3)	96 (11.8)	284 (34.7)	65 (7.9)

processing activities offshore. However, as a destination for Taiwanese (and, for that matter, Hong Kong) FDI, China possesses yet a third location advantage, which is primarily due to a similar cultural and linguistic background that effectively lowers the transaction costs of doing business there. This last feature may prove to be all the more important for FDI firms with meager ownership advantage. Theory suggests that the existence of an ownership advantage provides the ultimate basis for an FDI undertaking.[2] The firm making FDI must be competitive enough to earn a profit in the foreign country, where trans-

Recipient and date	*Investor*				
	World total	*Hong Kong*	*Taiwan*	*Japan*	*United States*
Malaysia (continued)					
1988	1,863 (100)	114 (6.1)	317 (17.0)	467 (25.1)	204 (11.0)
1989	3,194 (100)	130 (4.1)	797 (25.0)	993 (31.1)	119 (3.7)
1990	6,517 (100)	139 (2.1)	2,345 (36.0)	1,557 (23.9)	210 (3.2)
1991	6,202 (100)	218 (3.5)	1,312 (21.1)	1,348 (21.7)	654 (10.5)
1992	6,977 (100)	31 (0.4)	589 (8.4)	1,504 (15.1)	1,295 (18.6)
1993	2,444 (100)	36 (1.5)	346 (14.2)	608 (24.9)	674 (27.6)
1994	3,826 (100)	350 (9.1)	1,150 (30.1)	706 (16.5)	501 (13.1)
1986–94	32,378 (100)	1,075 (3.3)	6,958 (21.5)	7,512 (23.2)	3,743 (11.6)
Thailand					
1986	513 (100)	14 (2.8)	20 (3.9)	292 (57.0)	40 (7.8)
1987	1,055 (100)	42 (4.0)	160 (15.2)	583 (55.3)	69 (6.5)
1988	3,790 (100)	133 (3.5)	455 (12.0)	2,265 (59.7)	262 (6.9)
1989	3,987 (100)	206 (5.2)	517 (13.0)	1,979 (49.6)	182 (4.6)
1990	7,359 (100)	4,322 (58.7)	420 (5.7)	1,235 (16.8)	336 (4.6)
1991	2,447 (100)	107 (4.4)	317 (13.0)	949 (38.8)	400 (16.4)
1992	3,128 (100)	34 (1.1)	130 (4.2)	986 (31.5)	808 (25.8)
1993	4,293 (100)	0 (0.0)	215 (5.0)	2,705 (63.0)	431 (10.0)
1994	1,327 (100)	320 (24.1)	83 (6.3)	123 (9.3)	156 (11.8)
1986–94	25,643 (100)	5,372 (20.9)	2,151 (8.4)	8,719 (34.0)	3,061 (11.9)

Sources: FDI approval data adopted from host country official statistics: (Board of Investment for Thailand, Malaysian Industrial Development Authority for Malaysia, Capital Investment Coordinating Board (BKPM) for Indonesia, and Ministry of Foreign Trade and Economic Cooperation (MOFTEC) for China).

n.a. Not available.

a. January to October only.

action costs are typically higher than in the familiar home-country environment. The greater the ownership advantage a firm possesses, the more readily the higher costs of operating in a foreign country may be endured. This explains why such international giants as Coca-Cola, Nabisco, Volkswagen, Philips, Hitachi, and Mitsubishi were among the first to enter the Chinese market with FDI ventures. These firms were able to do so because they enjoyed a certain element of monopolistic power stemming from, for example, their established brand names and patent technologies.

During the early phases of Taiwanese investment on the mainland (1987–91) most FDI was carried out by small and medium-size enterprises (SMEs) with standardized technology and little worldwide reputation.[3] These firms seemed to possess two primary advantages: a knowledge of how to organize production efficiently and stable market access coming from their previous experience with original equipment manufacturing (OEM). Nevertheless, since recognized brand names and patent technologies were generally lacking at the firm level, these "ownership advantages" pertained more to a given industry as a whole than to each firm within the industry. This implies that the firms' FDI undertakings might have been facilitated in some way by other sources of advantage, the most obvious one apparently being the lower transaction costs stemming from cultural and linguistic proximity at the FDI location.[4] The fact that the number of instances of Taiwanese FDI in China is much greater, while the scale of projects is substantially smaller, than Taiwanese FDI in other low-wage countries (such as Malaysia and Thailand) seems to confirm that hypothesis. Moreover, since the ownership advantages of Taiwanese firms are typically confined to the realm of production and rarely extend to research and development (R&D) and marketing, the possibility of using other forms of transactions (for example, licensed production) is limited. As a result, the best way for these firms to recoup lost profit as domestic conditions worsen is to reorganize production in another location that is cost competitive—provided entry costs are low. In other words, internalization provides the only viable solution for these firms to salvage their eroding value-added at home.

Table 6-2 presents the results of a report on industrial distribution of Taiwanese FDI on the mainland that clearly shows the characteristics of the initial phase of investment.[5] One can clearly see that the rankings of Taiwanese investment by sector are significantly different from those of total FDI in China (which includes FDI from more advanced countries). In particular, Taiwanese FDI in transportation equipment and nonmetallic mineral products was minimal, whereas FDI in plastic products, processed foods, machinery, and miscellaneous products ranked considerably higher than average FDI from all sources. In total, Taiwanese realized (not just contracted) investment in the PRC amounted to $4.23 billion as of early 1993.

Outward FDI, especially that of a relocation type, will induce structural changes in the home and host countries, both in production and in

Table 6-2. *Industrial Distribution of World Total and Taiwanese FDI in China, Cumulative Realized Investment, 1979–92*
Amounts in millions of U.S. dollars

	World total FDI			*Taiwanese FDI*		
Manufacturing sector	*Amount*	*Rank*	*Percent share*	*Amount*	*Rank*	*Percent share*
Processed foods	2,494.17	7	6.11	395.02	3	9.33
Beverages and tobacco	693.74		1.70	46.65		1.10
Textile products	3,522.23	3	8.63	263.51	7	6.23
Wearing apparel	4,198.74	2	10.28	299.82	5	7.08
Leather, fur, and articles thereof	1,501.98		3.68	153.72		3.63
Wood, bamboo, and rattan products	18.33		2.98	223.11		5.27
Pulp, paper products, and printed matter	2,372.43	8	5.81	228.74	10	5.40
Chemicals	3,152.37	4	7.72	344.49	4	8.14
Chemical products	562.72		1.38	26.53		0.63
Oil and coal products	663.10		1.62	3.62		0.09
Rubber products	698.19		1.71	90.84		2.15
Plastic products	2,802.42	5	6.86	417.04	2	9.85
Nonmetallic mineral products	2,752.43	6	6.74	191.53		4.52
Basic metals	649.66		1.59	22.87		0.54
Metal products	1,983.21	9	4.86	242.89	8	5.74
Machinery	1,700.08		4.16	265.22	6	6.27
Electrical machinery and apparatus	6,172.12	1	15.12	555.51	1	13.12
Transportation equipment	1,817.85	10	4.45	164.79		3.89
Precision instruments and equipment	649.59		1.59	56.44		1.33
Miscellaneous products	1,228.01		3.01	240.38	9	5.68
All manufactures	40,833.38		100.00	4,232.73		100.00

Source: Charng Kao and Shih-ing Wu, "An Investigation of Foreign Direct Investment in Mainland China" (in Chinese), project report commissioned by the Ministry of Economic Affairs, Investment Commission, Republic of China, 1994.

trade. It is possible to estimate these impacts on the home and host economies by using input-output analysis.[6] A simplified version of such an analysis is summarized in equations 6-1 to 6-4, which display the main relationships among outward FDI, the generation of host-country output and exports, the derived demand for intermediate inputs, and the FDI-induced "reverse imports" from the host to the home economy.[7] Equation 6-1 shows the increase in output in the PRC (ΔY_i^*) as a function of the amount of FDI (ΔK_i) and the overseas output-capital ratio or capital productivity $[(Y_i/K_i)^*]$. This corresponds to that part of the output that is redeployed from home.[8] Equation 6-2 shows the FDI-induced increase in exports from China (ΔX_i^*) to the rest of the world, where a_i is the export propensity of the FDI firms in the ith industry. Equation 6-3 shows the derived demand for intermediate goods from Taiwan (ΔT, a 29 × 1 vector) created by FDI-related production in China (ΔY^*, a 29 × 1 vector), where D is the 29 × 29 domestic input coefficient matrix derived from the 1989 input-output table for Taiwan, and d_i is a sector-specific scaler representing the propensity to acquire these inputs from original sources by the ith-sector FDI firms.[9] Equation 6-4 shows the reverse imports, or sellbacks, to the home market of the ith industry (M_i), based on an estimate of g_i, the reverse-import propensity of the ith-sector FDI firms operating in China.[10]

(6-1) $\Delta Y^*{}_i = DK_i(Y/K)_i^*$ I = 1. 29

(6-2) $\Delta X^*{}_i = a_i\, DY_i^*$ I = 1. 29

(6-3) $\Delta T = d_i\, \Delta DY^*$ I = 1. 29

(6-4) $\Delta M_i = g_i\, Y_i^*$ I = 1. 29

The results of this exercise are shown in table 6-3. From a cumulative FDI of $4.23 billion (taken from column 3 of table 6-2), an estimated $13.95 billion of overseas production was generated in 1992, equal to roughly 7.3 percent of Taiwan's domestic manufacturing output for the same year. Of this output about $9.3 billion was exported to the world market from China, accounting for 11.6 percent of China's total exports in 1992. To manufacture this output, however, an estimated $5.9 billion of intermediated goods were imported from Taiwan to China, equal to roughly 7.2 percent of Taiwan's total exports in 1992. Finally, an estimated $2.42 billion of imports from China to Taiwan were created in the form of FDI sellbacks, making up about 4.2 percent of total imports to Taiwan in 1992.[11]

It has been argued that Taiwan, by relocating production to China, is duplicating the Japanese strategy (in the 1970s and 1980s) of indirectly exporting to the United States and other advanced-country markets.[12] Basically, however, successive waves of Taiwanese FDI toward China were, more than anything, a manifestation of the changing comparative advantages in production and trade between the two economies. In retrospect, one can easily discern this shifting comparative advantage between Taiwan and China through their recent export performance (tables 6-4 and 6-5). Using a constant-trend growth model, one can

Table 6-3. *Estimates of Taiwanese FDI Output, Exports, and Sellbacks vis-à-vis China, 1992*

Millions of U.S. dollars unless otherwise indicated

		Output			*Exports*					
Manufacturing sector	*Sectoral FDI*	*Estimated FDI*	*Taiwan's domestic*	*FDI as percent of domestic*	*Estimated FDI*	*China's total*	*FDI as percent of total*	*Estimated FDI sellbacks*	*Taiwan's total imports*	*Sellbacks as percent of total imports*
Processed foods	395.02	1,090.26	15,673.12	6.96	745.30	2,682.12	27.79	4.03	2,045.38	0.20
Beverages and tobacco	46.65	150.21	5,063.30	2.97	105.15	771.27	13.63	0.00	390.39	0.00
Textile products	263.51	793.17	11,720.00	6.77	432.83	5,941.99	7.82	95.18	1,499.44	6.35
Wearing apparel	299.82	902.46	6,375.95	14.15	804.09	16,237.66	4.95	45.84	537.51	8.53
Leather products	153.72	905.41	2,474.54	36.59	761.72	3,973.05	19.17	5.79	370.94	1.56
Wood products	223.11	698.33	2,821.69	24.75	675.71	1,820.67	37.11	196.65	977.70	20.11
Paper and paper products	228.74	523.81	8,783.18	5.96	234.62	524.64	44.72	167.41	1,689.05	9.91
Chemicals	344.49	809.55	14,326.44	5.65	84.92	2,482.64	3.42	431.73	5,790.75	7.46
Chemical products	26.53	78.00	4,805.85	1.62	14.00	1,639.93	5.18	8.18	3,285.89	0.25
Oil and coal products	3.62	6.91	9,663.69	0.63	5.99	4,692.35	0.13	0.69	4,459.41	0.01
Rubber products	90.84	274.33	2,381.30	11.52	243.14	312.57	77.79	27.43	n.a.	n.a.
Plastic products	417.04	1,084.20	11,525.74	9.41	715.86	3,637.46	19.68	340.90	884.98	38.52
Nonmetallic mineral products	191.53	296.87	6,293.91	4.72	158.78	3,134.87	5.06	9.17	792.57	1.16
Basic metals	22.87	36.82	13,401.69	0.27	12.40	3,366.34	0.37	0.00	8,625.27	0.00
Metal products	242.89	699.52	10,055.72	6.96	211.75	1,184.64	17.87	172.85	820.27	21.07
Machinery	265.22	745.27	8,309.51	8.97	562.75	3,366.82	16.71	533.39	6,988.04	7.63
Electrical machinery	555.51	3,044.19	33,074.03	9.20	1,990.29	8,175.21	24.35	213.76	9,307.76	2.30
Transportation equipment	164.79	479.54	15,189.55	3.16	384.35	2,203.29	17.44	7.24	6,044.98	0.12
Precision instruments	56.44	309.29	2,199.06	14.06	126.96	2,258.84	5.62	21.71	2,141.35	1.01
Miscellaneous products	240.38	1,024.02	6,690.10	15.31	1,005.38	11,256.78	8.93	137.94	755.33	18.26
All manufactures	4,232.73	13,952.26	190,828.37	7.31	9,275.99	79,663.14	11.64	2,419.89	57,417.44	4.21

Sources: *The Input-Output Table of Taiwan*, 1989; *Monthly Statistics of Import/Export Trade, Taiwan Area* (Taipei: Ministry of Finance, Dept. of Statistics, various issues); *Industrial Production Statistics Monthly, Taiwan Area* (Taipei: Ministry of Finance, Dept. of Statistics, various issues); *China Customs Statistics* (Hong Kong: Economic Information Agency, various issues); and FDI figures from table 6-2.

measure this shifting comparative advantage ex post as the difference between actual export growth and constant-trend export growth. I chose the year 1986 as the dividing year marking the beginning of macroeconomic changes in the productive environment in Taiwan (see chapter 3). With Taiwan's and China's 1986 exports as the basis for observation, the constant-trend exports for the two economies in 1992 are calculated by using the average growth rate of the period 1982–86 for each country as the constant-trend growth rate. The difference between 1986 exports and expected 1992 exports, given an unchanged growth rate, is the constant-trend export gains (column 5 of tables 6-4 and 6-5). But if one subtracts the 1986 actual exports from the 1992 actual exports, one obtains the 1992 actual export gains (column 6 of the tables). The difference between columns 5 and 6 (or, alternatively, between columns 3 and 4) may be viewed as an ex post indicator of the extent of the shift in comparative advantages for each economy between 1986 and 1992. From an ex ante point of view, all sectors with an asterisk in tables 6-4 and 6-5 are expected to show the clearest shift in comparative advantage.[13] These include the manufacture of apparel, leather and fur products, wood and bamboo products, miscellaneous products (such as toys and sporting goods), plastic products, nonmetallic mineral products, metal products, and electrical and electronics products. These sectors all fall into the labor-intensive category that met with difficulties in Taiwan from 1986 to 1992.

These expectations are supported by my results in table 6-4, in which all those sectors are shown to have a negative sign in the difference between actual and trend exports for Taiwan. As for the other sectors, the calculated difference between actual and trend exports are all positive, indicating a shift in comparative advantage that was favorable for the production and export of these goods in Taiwan (though the magnitudes involved tend to be small). Loosely speaking, from 1986 to 1992 Taiwan's traditional labor-intensive sectors "lost" some $35 billion worth of exports because of shifting comparative advantages, whereas the export "gains" from the other sectors amounted to only $5.6 billion, rendering a net "loss" of $29.4 billion in exports for Taiwan.

The situation is the reverse for China. All the labor-intensive sectors except for wearing apparel registered positive differences between actual and trend exports from 1986 to 1992 (table 6-5). The magnitudes of these gains range from $0.5 billion to $7.1 billion, implying that after 1986 the shift in international comparative advantage in producing these labor-intensive items tilted in favor of China at the expense of

Taiwan. According to my calculations, this change in comparative advantage already created a cumulative gain of $32.5 billion in exports for China (as of 1992). Furthermore, one finds an almost all-out positive gain across all manufacturing sectors for China. This stands as a sharp contrast to the Taiwan situation, in which some sectors gained at the expense of others. The asymmetry in performance can have important policy implications, because it suggests that the shift in comparative advantage between the two economies may be comprehensive rather than apply only to individual sectors.

By the mid-1990s, a decade after Taiwanese SMEs first went to China, Taiwanese FDI has entered a qualitatively new phase, both in scale and orientation. The average size of investment into China rose sharply from $735,000 in 1991 to $2.78 million in 1995. Firms making the investments are now publicly listed companies rather than small and medium-size firms. The motivation for FDI operations has turned from an export orientation toward local-market exploitation. The locale of FDI has shifted from the south farther to the north in China, and the areas of investment are assuming ever-stronger capital and technology intensity.[14] Table 6-6 provides a glimpse of Taiwan's publicly listed companies that have established a presence in China as of mid-1996. It is impressive that FDI is being made across the board, covering all industrial categories, since Taiwanese publicly listed companies were not even allowed to invest in China until as recently as 1990.[15] These developments seem to lend strong support to the hypothesis that improvement in China's comparative advantage over Taiwan has been comprehensive rather than sporadic.

If that is true, there may be grave implications for the possible future course of division of labor between Taiwan and China. For Taiwanese businesses, China has three types of advantage as a production base: low costs, big market, and minimal entry barriers. When political barriers between Taiwan and China abate, large-scale migration of industrial activities will most likely become rational, and even optimal, because the size of the China market is many times that of Taiwan, and operation costs are several times cheaper. What kind of division of labor will then prevail within the manufacturing sector will become a critical question for policymakers in Taiwan. Is Taiwan, for example, ready for a division of labor that is given by a "corner solution" (that is, total specialization) in which Taiwan concentrates its resources on, for instance, providing "local" services? These and related questions may be among the toughest to answer by students of economic changes wit-

Table 6-4. *Taiwan's Actual and "Constant-Trend" Export Growth, 1986–92*

Millions of U.S. dollars unless otherwise indicated

Manufacturing sector	*1986 actual exports (1)*	*1982–86 average growth rate (percent) (2)*	*1992 trend exports (1) × (2) (3)*	*1992 actual exports (4)*	*Difference in trend exports, 1986–92 (3) – (1) (5)*	*Difference in actual exports, 1986–92 (4) – (1) (6)*	*Difference between actual and trend exports (6) – (5) (7)*
Processed foods	1,934	12.77	3,978	2,606	2,044	672	–1,372
Beverages and tobacco	14	–4.17	11	49	–3	35	38
Textile products	2,636	14.24	5,859	6,722	3,223	4,086	863
Wearing apparel	4,663	9.51	8,042	4,572	3,379	–91	–3,470[a]
Leather products	1,287	25.27	4,973	1,081	3,686	–206	–3,892[a]
Wood and bamboo products	1,762	12.22	3,519	1,070	1,757	–692	–2,449[a]
Miscellaneous products	4,287	13.25	9,044	5,098	4,757	811	–3,946[a]
Nondurables subtotal	16,583	13.49	35,426	21,198	18,843	4,615	–14,228
Paper and paper products	250	11.32	476	827	226	577	351
Chemicals	782	16.21	1,926	2,498	1,144	1,716	572
Chemical products	698	4.56	912	1,448	214	750	536
Rubber and plastic products	3,934	22.26	13,168	5,122	9,225	1,179	–8,046[a]
Nonmetallic mineral products	791	13.15	1,660	1,217	869	426	–443
Basic metals	702	1.85	784	1,453	82	751	669[a]
Metal products	2,360	23.36	8,317	5,304	5,957	2,944	–3,013[a]
Intermediate subtotal	9,527	19.14	27,243	17,869	17,716	8,342	–9.373

Manufacturing sector	*1986 actual exports (1)*	*1982–86 average growth rate (percent) (2)*	*1992 trend exports (1) × (2) (3)*	*1992 actual exports (4)*	*Difference in trend exports, 1986–92 (3) – (1) (5)*	*Difference in actual exports, 1986–92 (4) – (1) (6)*	*Difference between actual and trend exports (6) – (5) (7)*
Machinery	1,598	17.27	4,156	5,731	2,558	4,133	1,575
Electronics and electrical products	8,910	22.85	30,629	22,222	21,719	13,312	–8,407[a]
Transportation equipment	1,749	12.49	3,544	4,049	1,795	2,300	505
Precision instruments	740	11.60	1,430	1,925	690	1,185	495
Durable and capital goods subtotal	12,997	20.48	39,759	33,927	26,762	20,930	–5,832
All manufactures	39,107	17.41	102,428	72,995	63,321	33,888	–29,433

Sources: *The Input-Output Table of Taiwan*, 1989; *Monthly Statistics of Import/Export Trade, Taiwan Area* (Taipei: Ministry of Finance, Dept. of Statistics, various issues); *Industrial Production Statistics Monthly, Taiwan Area* (Taipei: Ministry of Finance, Dept. of Statistics, various issues); *China Customs Statistics* (Hong Kong: Economic Information Agency, various issues); and FDI figures from table 6-2.

a. Indicates sectors with a significant shift in comparative advantage.

Table 6-5. *China's Actual and "Constant-Trend" Export Growth, 1986–92*

Millions of U.S. dollars

Manufacturing sector	*1986 actual exports (1)*	*1982–86 average growth rate (percent) (2)*	*1992 trend exports (1) × (2) (3)*	*1992 actual exports (4)*	*Difference in trend exports, 1986–92 (3) – (1) (5)*	*Difference in actual exports, 1986–92 (4) – (1) (6)*	*Difference between actual and trend exports (6) – (5) (7)*
Processed foods	1,907	24.15	6,983	2,682	5,076	775	–4,301
Beverages and tobacco	120	5.55	166	771	46	651	605
Textile products	3,306	1.74	3,667	5,942	361	2.636	2,275
Wearing apparel	5,110	27.28	21,726	16,238	16,616	11,128	–5,488[a]
Leather products	227	9.75	397	3,973	170	3,746	3,576[a]
Wood and bamboo products	168	–2.16	147	1,821	–21	1,653	1,674[a]
Miscellaneous products	1,398	6.85	2,080	11,257	682	9,859	9,177[a]
Nondurables subtotal	12,236	19.24	35,166	42,684	22,930	30,448	7,518
Paper and paper products	200	8.56	327	525	127	325	198
Chemicals	801	3.86	1,005	2,483	204	1,682	1,478
Chemical products	897	17.01	2,302	1,640	1,405	743	–662
Rubber and plastic products	391	4.82	519	3,950	128	3,559	3,431[a]
Nonmetallic mineral products	322	–0.15	319	3,135	–3	2,813	2,816
Basic metals	670	0.29	682	3,366	12	2,696	2,684[a]
Metal products	563	1.87	629	1,185	66	622	556[a]
Intermediates subtotal	3,844	7.04	5,783	16,284	1,939	12,440	10,501

Manufacturing sector	*1986 actual exports (1)*	*1982–86 average growth rate (percent) (2)*	*1992 trend exports (1) × (2) (3)*	*1992 actual exports (4)*	*Difference in trend exports, 1986–92 (3) – (1) (5)*	*Difference in actual exports, 1986–92 (4) – (1) (6)*	*Difference between actual and trend exports (6) – (5) (7)*
Machinery	381	–10.24	199	3,367	–182	2,986	3,168
Electronics and electrical products	487	14.28	1,085	8,175	598	7,688	7,090[a]
Transportation equipment	255	–11.18	125	2,203	–130	1,948	2,078
Precision instruments	151	–1.14	141	2,259	–10	2,108	2,118
Durables and capital goods subtotal	1,274	3.32	1,550	16,004	276	14,730	14,454
All manufactures	17,353	16.10	42,499	74,971	25,145	57,618	32,473

Sources: *The Input-Output Table of Taiwan*, 1989; *Monthly Statistics of Import/Export Trade, Taiwan Area* (Taipei: Ministry of Finance, Dept. of Statistics, various issues); *Industrial Production Statistics Monthly, Taiwan Area* (Taipei: Ministry of Finance, Dept. of Statistics, various issues); *China Customs Statistics* (Hong Kong: Economic Information Agency, various issues); and FDI figures from table 6-2.

a. Indicates sectors with a significant shift in comparative advantage.

Table 6-6. *Number and Share of Taiwan's Publicly Listed Companies Investing in China, by Industry*

		Firms investing in China	
Industry	*Total number of firms*	*Number*[a]	*Percent of total*
Cement	9	3	33.33
Processed food	28	13	46.43
Plastic materials	18	5	27.78
Textiles	46	8	17.39
Electrical products	17	5	29.41
Electronics	48	12	25.00
Cables and wires	12	5	41.67
Petrochemicals	19	7	36.84
Ceramics and porcelain	6	3	50.00
Paper products	7	1	14.29
Steelmaking	24	1	4.17
Rubber products	8	4	50.00
Automobiles	5	1	20.00
Transportation	14	1	7.14
Construction	36	1	2.78
Tourism	6	0	0.00
Total	358	83	23.18

Source: Ministry of Finance, Taiwan Securities Exchange Commission, Republic of China.
a. Statistics are as of mid-1996; they do not include firms investing through individual channels.

nessing rapidly shifting comparative advantage across different regions with different levels of economic achievement. The process of factor-price equalization is now accelerating and being reinforced by the free mobility of capital and managerial resources,[16] countering endeavors by governments around the world to maintain their positions relative to one another. This process is likely to be further intensified by the emergence of China on the world's economic landscape.

Sectoral Analysis of the Electronics Industry

Electronics has been a major field for Taiwanese investment in the PRC since 1987. As with other lines of production, outward FDI in electronics has also engendered a substitution of production and exports between the two economies. Here I take a closer look at the electronics sector in Taiwan, with particular emphasis on the personal computer industry, to shed more light on the emerging division of labor across the Taiwan Strait, as well as on some of the potential problems and policy issues surrounding this trend. The PC industry is significant for two reasons. First, it is a rare sunrise industry in the Taiwan economy, one that has enjoyed continuous buoyant growth, even after the mid-1980s. Second, it is the only manufacturing sector in Taiwan whose growth performance has not been related to the emergence of the Chinese economy either as a nearby export market or as a regional production base—at least until very recently. Taiwan's PC-related exports were shipped to all the major external markets, but very little went to explore the Chinese inland markets until after the 1990s. The same was true for FDI activities. However, once the process was triggered, a quick-paced division of labor between Taiwan and China emerged, as did significant intersectoral and intrasectoral shifts in both the investing and the invested country. It is therefore of particular interest, from the perspectives of both economies, to study the impact and implications of the rapidly enhanced interaction between Taiwan and China in this important sector of the Taiwan economy.

To summarize briefly, Taiwan's electronics industry was initiated by an influx of foreign direct investment from Western and Japanese multinational corporations (MNCs). Firms like RCA, Zenith, Matsushita, and Netscape Communications (NEC) came to Taiwan in the 1960s and 1970s to set up wholly owned subsidiaries or local joint ventures for the production of transistor radios, black-and-white television sets, and electronic components and parts. Even though such investment seldom exceeded 25 percent of total sectoral capital formation, foreign-affiliated production accounted for more than 60 percent of the island's electronics exports in the 1970s. Later on, however, domestic entrepreneurs successfully emulated the local MNCs in setting up their own assembly lines and gradually became the center of gravity of Taiwan's electronics production. The sector as a whole expanded steadily during the 1970s, overtaking textiles by the early 1980s to become the biggest foreign exchange earner in the Taiwan economy.

With the advent of macroeconomic changes in the latter half of the 1980s, the electronics industry in Taiwan underwent a major restructuring of its product composition. Domestic output of radio receivers, tape recorders, television sets, and electrical household appliances declined sharply as exports of these items were rapidly replaced by competing products from China and other developing countries (table 6-7). Part of

Table 6-7. *Production Index for Subsectors of the Electronics Industry in Taiwan, 1984–95*

1991 = 100

Year	*TV set and video tape recorder and player*	*Radio tape recorder and record player*	*Record and stereo equipment parts*	*Electric fan*	*Lighting fixture*	*Electric heating appliances*	*Other electric appliances and house-ware*
1984	80.83	247.65	65.11	116.16	128.72	49.11	28.02
1985	80.65	209.53	59.47	121.21	131.46	51.10	24.65
1986	119.10	225.28	77.48	144.62	128.25	65.37	29.87
1987	159.13	247.84	94.02	148.49	123.05	75.76	41.40
1988	173.91	195.44	95.78	131.74	140.10	89.13	60.61
1989	161.86	155.57	98.54	97.46	110.58	90.75	84.16
1990	93.84	113.05	95.92	85.48	98.56	83.19	76.40
1991	100.00	100.00	100.00	100.00	100.00	100.00	100.00
1992	79.33	80.50	76.94	89.21	81.66	90.74	92.41
1993	63.76	66.94	68.36	64.26	61.84	77.76	83.37

Year	*Data storage media unit*	*Data terminal equip-ment*	*Data I/O peripheral equipment*	*Computer compon-ents*	*Other computer equipment*	*Electronic tube and semicon-ductors*	*Other electronic parts and com-ponents*
1984	16.63	38.79	44.87	14.42	19.47	48.41	33.19
1985	20.27	41.48	42.08	20.64	23.74	39.39	32.32
1986	29.90	58.45	43.28	34.13	35.39	55.64	42.92
1987	43.38	73.61	60.31	56.21	52.44	64.97	55.54
1988	39.17	89.79	64.61	51.66	49.25	72.24	67.77
1989	48.99	87.26	62.70	70.36	61.31	78.26	71.13
1990	84.54	94.58	87.00	84.11	82.88	93.02	84.53
1991	100.00	100.00	100.00	100.00	100.00	100.00	100.00
1992	146.82	116.31	117.92	110.81	114.44	123.80	109.30
1993	187.15	125.61	135.48	125.39	123.60	147.27	131.81
1994	205.29	136.66	147.72	179.93	120.45	146.95	166.48
1995	234.11	168.93	146.00	251.04	166.51	146.23	216.11

Source: *Monthly Statistics of Industrial Production of Taiwan* (Taipei: Dept. of Statistics, Ministry of Economic Affairs, various issues).

the domestic production has since been relocated overseas, and the rest simply faded away. Thanks to the timely and accelerated growth of the PC industry, the gap created by the decline in older products was filled, and the electronics sector as a whole grew strongly into the 1990s. By 1995 its output level reached $60 billion, accounting for 23 percent of Taiwan's manufacturing GDP and over 35 percent of its total exports.[17]

The PC industry, unlike the earlier consumer electronics industries, was established in Taiwan largely by indigenous effort.[18] The preceding television and calculator industries had bestowed on Taiwan a well-rounded local component industry and a proficiency in the manufacture and assembly of audio and video equipment, thereby providing a solid foundation for the subsequent development of PC products such as terminals, monitors, and other peripheral items. In the late 1970s and early 1980s a sudden boom in the manufacture of videogame machines and Apple II clones further generated a large number of local engineers experienced in circuit design and the application of microprocessors.[19] The concurrent establishment of the Institute for Information Industry (III) in Taipei, the Electronics Research Service Organization (ERSO), and the Science-Based Industrial Park in Hsinchu, along with favorable investment incentives accorded by the government, attracted both local talent and a significant backflow of overseas Chinese with computer technology backgrounds to set up PC-related businesses on the island. In 1983, one year after the world's first IBM-compatible PC XT was introduced by Compaq, Acer (then already a local champion of Chinese I/O terminals and education PCs) presented Taiwan's first IBM-compatible PC XT and ignited the subsequent rapid expansion of the industry.[20] Total PC-related output grew to an impressive $15.8 billion in 1995, 90 percent of which being furnished by indigenous firms.[21] Together with an offshore production value of $5.4 billion, Taiwan surpassed Germany to become the world's third-largest PC maker in 1995, surpassed only by the United States and Japan. Among Taiwan's major product items are desktop and portable PCs, computer monitors, motherboards, keyboards, PC mice, and switch power supply units. Many of these items accounted for more than half the world supply (table 6-8).

Several distinctive features of the Taiwanese PC industry are of interest here. First, it has a very high export propensity: 63 percent of total output in 1995. High export propensity implies intense pressure from international competition. As a result, optimizing productive efficiency and reducing operational costs are of primary concern to firms. Second, Taiwan has established an extensive network of vertical linkages within the industry (though not necessarily within each firm), ranging from upstream electronic components and parts (logic integrated circuits, memory IC, chip sets, smaller liquid crystal displays

Table 6-8. *Domestic and Overseas Production of Taiwan's Top Twelve PC Products, 1995*
Production in 1,000 sets

Rank	*Product item*	*World market share (percent)*	*Domestic production*	*Annual growth rate (percent)*	*Overseas production*	*Growth rate (percent)*	*Overseas/total production ratio (percent)*
1	Monitor	57	16,085	12	15,244	58	49
2	Personal computer	12	6,758	39	400		n.a.
	Desktop	10	4,167	48	400	n.a.	9
	Portable	27	2,592	26	n.a.	n.a.	n.a.
3	Motherboard	65	13,133	14	7,751	29	37
4	Switch power supply	35	7,756	261	26,564	12	77
5	Image scanner	64	2,481	49	n.a.	n.a.	n.a.
6	Graphics card	32	4,920	–2	4,380	17	47
7	Keyboard	65	4,589	–35	28,191	79	86
8	CD-ROM drive	11	2,825	1,519	927	927	25
9	Network interface	38	9,946	63	318	1,490	3
10	Terminal	27	956	–8	n.a.	n.a.	n.a.
11	Audio card	35	1,663	–16	n.a.	n.a.	n.a.
12	Mouse	72	31,087	41	9,817	27	24

Source: Institute for Information Industry, Market Intelligence Center, Republic of China. Hereafter III-MIC.
n.a. Not available.

[LCDs], cathode ray tubes [CRTs], and motherboards), midstream peripheral items (keyboards, monitors, image scanners, PC mice, and power supply units), to the final assembly of desktop and portable PCs. The closely integrated local production network, together with a crucial dose of domestic design capacity (developed mainly through publicly supported institutions, such as ERSO and III, but also within private firms) helped form the OEM and ODM (original design manufacture) credibility of Taiwan, making it one of the world's leading manufacturing centers of PC products. Third, Taiwanese PC firms are renowned for their fast speed in catching up with the most recent development in the marketplace and positioning themselves accordingly as accommodating suppliers of PC-related products.[22] Thus with the advent of network-related applications, Taiwanese SME PC producers were among the first to grab market niches and furnish multimedia audio and video cards, image scanners, modems, switches, bridges, and CD-ROM drives. These firms often became significant suppliers within a short time.[23] Fourth, while many small firms were emerging to take advantage of the swiftly changing technology and new market niches, successful older firms were able to grow in scale with the help of fierce market competition. Take the largest subsector—monitors—for example. The ten biggest producers in Taiwan accounted for more than 70 percent of production in 1992, and the trend is for continuing expansion of company scale. With increasing scale, these firms have been able to cut down on average cost and gain further competitiveness.

However, there are also weaknesses in the Taiwan PC industry. First of all, Taiwanese PC producers are still too limited in size by international standards. Aside from Acer, Tatung, and Mitac, no Taiwanese PC producers are on the list of the world 100 PC producers. Because of their small size, these firms are financially weak (and often less motivated) to support major R&D activities and rarely engage in self-conducted marketing. Most of them rely on OEM-ODM orders to supply PCs to the international market in IBM-compatible models. More important, Taiwanese firms are still unable to supply for themselves some of the key components of PCs (such as central processing units [CPUs] and, until recently, dynamic random-access memory [DRAMs]) and depend heavily on the United States and Japan (and, in some cases, Korea) for these items. Although its upstream integrated circuit industry spent some 6 percent of its annual turnover on R&D activities in 1995, that amounted to only one-fifth of the R&D expenditure made by Intel alone.[24] Except for a few prestigious firms that are able to cross-license technology with leading multinational firms (such as Macronix with IBM), most Taiwanese companies are basically followers, with minor modifications, of the latest product designs developed elsewhere. In this sense the PC industry in Taiwan is still largely an assembly-type manufacturing activity, with more than 60 percent of its exports being OEM-ODM production not associated with Taiwan producers' own brand names.[25] As a result, although total sales volumes are substantial, the profits these firms get out of their operations and the value-added they provide tend to be small.

FDI by Taiwanese PC Firms

As with other industries on the island, Taiwan's PC industry in recent years also suffered from a deteriorating investment climate at home. An acute labor shortage and escalating domestic land costs inevitably eroded Taiwan's comparative advantage in producing lower-end PC products. Before the China option was open to Taiwanese firms, most of them were forced to relocate their labor-intensive production processes to ASEAN countries, especially Malaysia and Thailand. Figure 6-1 traces the timeline of Taiwanese hardware producers' FDI pilgrimage toward low-wage countries. It is interesting to note the sequence in which different products are moved overseas, starting from the most labor-intensive keyboards and PC mice to switch power supply units, and then to motherboards and monitors. Leading peripheral producers such as Acer Peripherals, Silitek, and Clevo led the way in

Figure 6-1. *Timeline of Taiwanese PC Producers' Outward Foreign Direct Investment, 1988–96*

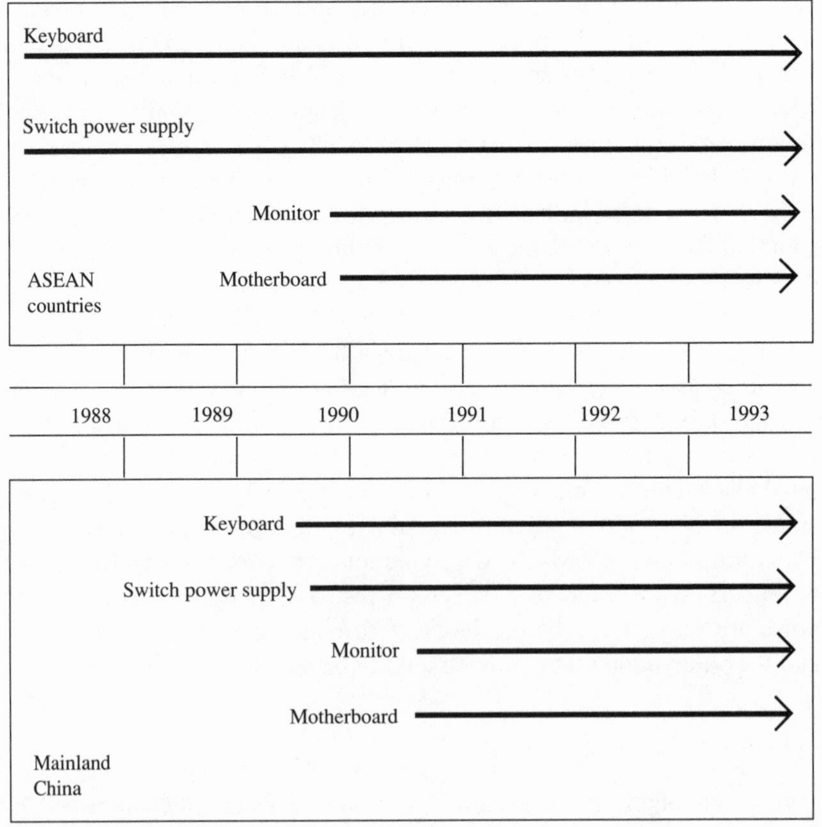

Table 6-9. *Taiwan's Major PC-Related Projects in Malaysia, Thailand, and China*

Destination	*Producer and content of investment*
Malaysia	Acer Peripherals (keyboard, monitor), Lite-On Technology (monitor) Silitek (keyboard), Clevo (keyboard), Shamrock (monitor), Taiwan Liton (switch power supply), Rectron (switch power supply)
Thailand	Tatung (monitor), Analog Devices (monitor), Chuntex (monitor), Delta Electronics (switch power supply), Compal (monitor), Chicony Electronics (keyboard, motherboard), Auto-Computer (motherboard, keyboard), Capetronic (monitor)
Mainland China	Acer Peripherals (monitor, keyboard, motherboard), First International Computer (motherboard, monitor), Copam Electronics (monitor), Mitac International (monitor, PC assembly), Datatech Enterprises (motherboard), Logitech (PC mouse), Primax Electronics (PC mouse), Chung Hua Picture Tubes (CRTs), Picvue Electronics (small-size LCDs)

Sources: *Data Bank on the Production and Sales of the Electronics and Information Industry* (in Chinese), III-MIC, 1996, and various press releases.

relocating keyboard assembly lines to Southeast Asia. They were soon followed by companies such as Taiwan Liton, Rectron, and Delta Electronics for the production of switch power supply units; Silitek, Chicony, and Autocomputer for the production of motherboards; and Tatung, Chuntex, and Lite-On Technology in the realm of monitors (table 6-9). Acer Peripherals, Autocomputer, and a number of other firms established extensive facilities in Southeast Asia that covered a wide range of PC-related products. The extent of overseas production in some of these items has been substantial. Take keyboards, for example. Almost 86 percent of the total output in 1995 was derived from overseas production. Similar figures for switch power supply units, monitors, motherboards, and PC mice were 77 percent, 49 percent, 37 percent, and 24 percent, respectively. Overall, offshore production of PC hardware already accounted for 27 percent of the industry output in 1995, having risen rapidly from 10 percent in 1992. Emerging products for home operation included CD-ROM drives and network interface, which were growing annually at 1,519 percent and 63 percent, respectively, and already accounted for 11 percent and 38 percent of world supply at the end of 1995.[26]

Figure 6-2. *Geographic Distribution of Taiwan's Offshore PC Hardware Production, 1992, 1993*

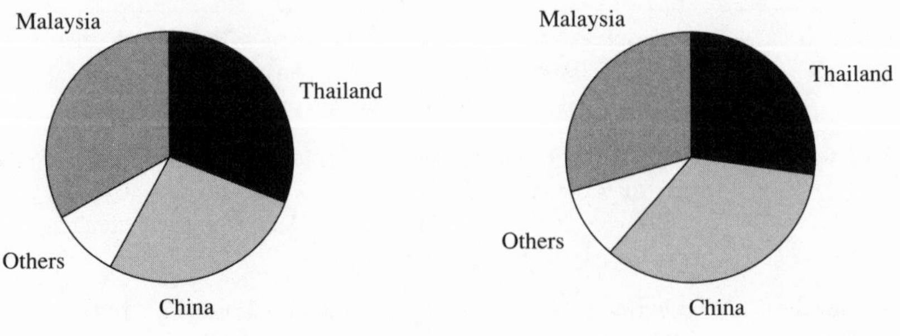

1992: Value of offshore production $787 million (excluding point-of-sale assembly $182 million)

1993: Value of offshore production $1,691 million

Source: III-MIC.

In the growing volume of offshore production, mainland China is claiming an increasing share. Taiwanese firms did not migrate across the Strait in large numbers until 1990.[27] By 1993, however, China's share of Taiwan's offshore production of PC hardware had already risen to 34.6 percent, surpassing both Malaysia's (29.4 percent) and Thailand's (27.3 percent) (figure 6-2). Although the ban on investment did not effectively prevent smaller firms from sneaking into China, it did have a prohibitive effect on the major (and thus more transparent) producers in Taiwan. When the government finally legalized investment in China for a certain number of PC-related products in 1992, leading producers such as Acer, First International Computer, and Mitac wasted no time in moving FDI into China. In 1993 there were already thirty-five Taiwanese PC subsidiaries in China, as against ten in Thailand, nine in Malaysia, and four in Indonesia (table 6-10).[28] The number increased to forty-one in 1995, constituting 70 percent of all PC firms that were running overseas subsidiaries.[29] China accounted for almost half of Taiwanese offshore production of motherboards in 1993, one of the latest items to go abroad, with the rest being supplied by subsidiaries based in Thailand and Malaysia. Similarly, operations in China produced some 2 million monitors in 1993, accounting for almost 80 percent of China's total monitor output and more than 50

Table 6-10. *Distribution of Taiwan PC Producers' Overseas Subsidiaries in Three ASEAN Countries and Mainland China*

	Number of establishments				
Product	*Mainland China*	*Thailand*	*Malaysia*	*Indonesia*	*Total*
Keyboard	5	2	3	2	12
Switch power supply	3	4	3	0	10
Monitor	9	4	3	2	18
Motherboard	7	2	0	1	10
Desktop PC	0	1	1	0	2
PC mouse	3	0	1	0	4
Drawing interface	8	0	0	0	8
Total	35	10	9	4	58

Sources: *Data Bank on the Production and Sales of the Electronics and Information Industry* (in Chinese); and III-MIC, 1996.

percent of the entire offshore production of monitors by Taiwanese firms.[30]

The cost factor is of course the main driving force underlying these developments. Table 6-11 compares the estimated production costs for major PC hardware components in Taiwan, the ASEAN countries, and China. Taking monitors as an example, it is estimated that operating on the mainland can save up to 3 percent on direct labor costs and another 5 percent on indirect costs (such as management and production overhead), so that total cost savings can amount to 8 percent of the final product value, which is much higher than the 5 percent savings expected by operations in the ASEAN countries. As the degree of labor intensity increases, total cost savings also increase; savings from production on the mainland rather than in Taiwan can be as high as 21 percent for keyboards and 22 percent for PC mice. Cost differentials of this magnitude prove important for Taiwanese firms, since most must rely on price competition in the world marketplace. Therefore, these differentials by themselves are enough to attract FDI into China at the expense of the ASEAN countries.[31]

Accelerating outward FDI by downstream firms also induces upstream producers to follow suit. Such CRT and LCD manufacturers as Chung Hua Picture Tubes and Picvue Electronics are quickly moving to China. Having already established two major facilities in Malaysia, Chung Hua Picture Tubes plowed more than $26 million into Suzhou and Shanghai for the production of 14-inch monochrome and

Table 6-11. *Estimated Cost Savings to Taiwan Companies of Manufacturing Various PC-Related Products in Overseas Operations, 1993*
Percent

Product and location	Material cost	Direct labor	Indirect cost	Total cost	Cost savings
Monitor					
Taiwan	85	4	11	100	...
ASEAN[a]	83	3	9	95	5
China	85	1	6	92	8
Motherboard					
Taiwan	80	5	15	100	...
ASEAN[a]	80	3	13	96	4
China	80	1	9	90	10
Switch power supply					
Taiwan	70	10	20	100	...
ASEAN[a]	69	5	18	92	8
China	70	2	12	84	16
Keyboard					
Taiwan	60	22	18	100	...
ASEAN[a]	56	13	15	84	16
China	56	10	13	79	21
Mouse					
Taiwan	52	23	25	100	...
China	48	10	20	78	22

Sources: *Data Bank on the Production and Sales of the Electronics and Information Industry* (in Chinese); and III-MIC 1996.

a. ASEAN refers mainly to Malaysia.

color CRTs, with additional investment plans laid out for 1995–96. Another major upstream producer that has showed great interest in China is Picvue Electronics, a leading manufacturer of LCDs in Taiwan. Picvue has already located a small LCD plant in Shanghai, hoping to supply the local PC industry as well as FDI firms originating from Taiwan. Similarly, Lite-on Technology, a major producer of diodes, announced two major investment projects in early 1996 to build factories in Tianjin and Guangzhou, respectively, for the production of diodes and related semiconductor components. A number of earlier entrants to the Chinese market, including WUS Printed Circuit Co. (printed circuit boards), Hon Hai Precision (calculator components), and Delta Electronics (power supplies and color monitors), are now among the most profitable electronics firms with a presence in China.

Judging from this persistent trend of FDI toward the PRC, close production cooperation combining Taiwanese capital, managerial skills, and OEM reputation with Chinese land and labor (both skilled and unskilled) seems an irreversible trend well into the future. In terms of division of labor across the Strait, the crucial question is whether Taiwanese firms are capable of moving up the product ladder and finding new niches for domestic operations, so that the PC industry may continue to thrive in Taiwan along with a successful redeployment of its lower-grade products to China.

Pressure on Domestic Restructuring

Theoretically, there are three possible ways in which a firm can deal with its domestic capacity after an overseas transplant is made. It can reduce the scale of domestic production or totally cease operations at home; engage in product differentiation and upgrading (horizontal integration); or diversify or vertically integrate, or both.

In Taiwan's case, one can distinguish between three kinds of PC producers, each choosing a different one of those options. The first type of firm, adopting the first option, consists of small and medium-size PC firms producing a single product on a limited scale. Because these firms were the hardest hit by the changing domestic environment in recent years, they were also the first to move offshore. Once they relocate production in China, they typically discontinue manufacturing activities in Taiwan, retaining only the partial functions of marketing and product development. Overseas production by these firms tends to concentrate on low-price, labor-intensive, and standard technology items such as keyboards, PC mice, and switch power supply units.

The second type consists of medium-size firms producing a spectrum of closely related products in Taiwan. They maintain operations at home through product upgrading and differentiation (that is, a horizontal division of labor between the parent and the subsidiary) but lack the managerial, financial, and technological capability to engage in vertical integration or production diversification. The majority of smaller-scale monitor and motherboard producers in Taiwan fall into this category. For example, when Mitac and Copam moved the assembly of small monitors to overseas bases, they switched to larger monitors for production at home. In 1993 production relocated in mainland China was mostly of 14-inch monochrome and color monitors, whereas domestic production was upgraded to higher-end products such as 15- and 17-

inch color super video graphics array (SVGA) monitors. Similarly, Clevo and Chicony switched their emphases to notebook PCs for home production soon after they moved their keyboard production to Malaysia and Thailand. Datatech also now concentrates on the production of desktop and portable PCs after moving its motherboard assembly lines to China.

The third type of firm consists of large-scale, multiproduct manufacturers that are increasingly approaching MNC status by international standards. Acer, First International Computer, and Tatung certainly belong to this select group. For these firms, overseas operations are an integral part of their global production and marketing strategy and therefore are viewed as a supplement to, rather than as a substitute for, domestic production. Another feature of these firms is that they were very cautious in moving into China. Acer, for example, was among the last to make a move, because of political concerns. But in the long run these are the firms most capable of extensive investments in the PRC. In 1994 major PC producers in Taiwan, led by Acer, persuaded the government to loosen restrictions on FDI ventures in China for the assembly of the post-486 (Pentium) PC series, which was then the most recent model introduced for production in Taiwan. The argument put forth was that the assembly process was labor intensive and thus better situated in China than in Taiwan. Given that, the question left for Taiwanese producers is one of domestic product realignment (for example, moving into the area of key PC components), which in turn will place these firms in direct competition with established foreign MNCs currently dominating the scene. Whether Taiwanese producers will be able to win this high-tech battle and gain a solid stance in the arena remains an open question.

Faced with this challenge, Taiwan has made a twofold response. First, at the government level, a regulation scheme of counterpart investment was put forth in 1992.[32] This requires certain (especially high-tech) China-bound FDI firms to commit to new investment projects at home in accord with the changing comparative advantage of the domestic economy. The government subsequently adopted a categorical regulation scheme to screen FDI projects destined for China, in which a positive list of more than 4,500 items are permitted as legal FDI toward the PRC and a negative list of a few military and government-sponsored items are not allowed. There is a gray area in between subject to special screening and guided by the counterpart investment principle. Apart from this categorical regulation scheme, any mainland-bound FDI project exceeding $10 million automatically falls into the special screening category. The main concerns here are with potential transplants of high-tech capacities that Taiwan is itself trying to develop, in the hope that Taiwan's current technological lead over China will not be disturbed by successive waves of Taiwanese FDI toward the mainland.

Second, at the firm level, major producers such as Acer, Tatung, and First International Computer took spontaneous steps toward diversifying their product range and, in particular, investing in vertically interrelated PC key components. Acer, for example, formed a joint venture with Texas Instruments (TI-Acer) in 1989 to set up Taiwan's first 6-inch wafer plant for the manufacturing of 4Mb DRAM. The facility was expanded (at a cost of $470 million) to produce 16Mb DRAM using 8-inch wafers in 1995, adding a further 20,000 wafers per month capacity. Since then it has spent another $1.3 billion for a new 8-inch wafer plant, which is expected to start production in 1997, for 64Mb DRAMs (and gradually extending into 256Mb DRAMs) at a geometry of 0.35/0.25 micron (table 6-12). Another major PC and peripherals producer in Taiwan, First International Computer (FIC), also extended its operations into the realm of semiconductor design and manufacturing, making high-powered 386 and 486 ICs in the early 1990s and multimedia ICs more recently. FIC is a member of the Formosa Plastics Group, the biggest enterprise group in Taiwan. Another spin-off from Formosa Plastics, Nan Ya Technology, also reached an agreement with OKI in 1994 to transfer 16/64Mb DRAM technology together with a joint venture to carry out production of these chips in Taiwan. The estimated total cost of the project amounted to $580 million, in addition to its new investment in the STN (switched telecommunications network) LCD plant at $44 million. A similar effort of production integration and upgrading is observed for Chung Hua Picture Tubes, a close affiliate of the consumer electronics giant Tatung. Since 1993 Chung Hua Picture Tubes has invested extensively in China, but it also has major upgrading plans for Taiwan in the areas of color STN LCDs (totaling $110 million) and large (32-inch) CRTs to be used in high-density television ($230 million), in addition to its newly acquired capacity for the production of 17-inch color video graphics arrays (VGAs).

Perhaps more illuminating are the waves of investment made by firms outside the PC industry proper—that is, those coming from the

Table 6-12. *Recent Taiwanese Investment in PC Key Components*

Company	*Content of investment*	*Amount (millions of U.S. dollars)*	*Monthly output (thousands)*	*Process technology*	*Expected start-up date*
Venguard International (plant IA and B)	8" DRAM, 4/16 Mb SRAM	1,380 n.a.	15, 15	0.5, 0.4	1994/4Q 1996/4Q
TI-Acer (plant I, IA, and plant II)[a]	6" DRAM, 4/16Mb, 8" DRAM, 16 Mb, 8" DRAM, 64 Mb	250, 470, 1,280	0, 20, 40	0.8/0.5 0.5 0.35	1992 1995/2Q 1997/2Q
TSMC[b] (plant III, IV, and V)	Foundry	2,900	35, 30, 25	0.5, 0.35	1995/3Q 1996–97
UMC[c] (plant III)	LOGIC, SRAM, foundry	910	25	0.5, 0.35	1995/3Q
UMC's 3 joint ventures	Foundry	3,030	25, 25, 25	0.35, 0.25	1996–97
Nan Ya	8" DRAM, 16/64Mb STN LCD	580 44	24 30	0.45	1996/4Q 1994/4Q
Powerchip	8" DRAM, 16Mb	75	25	0.4	1996/3Q
Mosel-Vitelic (plant II)	6" and 8" DRAM	1,450	25	0.55, 0.35	1997/1Q
SMV Technology (Siemens-Mosel JV)	64/256Mb DRAM, 1Gb DRAM	1,700	20	0.35, 0.25	1997–98
Winbond (plant III)	SRAM, Logic, DRAM	1,300	40	0.35, 0.25	1997/1Q
Macronix (plant II)	NV memory, ASIC	1,100	30	0.35, 0.25	1997/2Q
Ta Tung/Chung-hua Picture Tubes	Memory 17" and 32" CRT, color STN LCD	730 230 110	25 n.a. n.a.	n.a. n.a. Toshiba	1997 1994 1995
Teco	17" & 32" CRT	n.a.	n.a.	NEC	1997/3Q
Unitac Optical	Color TFT LCD	365	n.a.	For PC	1995–96
Prime View	Color TFT LCD	255	20	For PC	1997
Asia-Pacific Investment Co.	15", 17", and 32" CRT	n.a.	n.a.	n.a.	n.a.
Tysil Electronic Materials	Silicon wafer	400	n.a.	MEMC	1996/1Q
Nan Ya/Komatsu	Silicon wafer	500	n.a.	Komatsu	1997–98

Source: *Electronics Industry Yearbook in 1995* (Taipei: Dept. of Statistics, Ministry of Economic Affairs, 1996); and various news releases.

n.a. Not available.

a. Texas Instruments-Acer.

b. Taiwan Semiconductor Manufacturing Corporation.

c. United Microelectronics Corporation.

more traditional sectors of textiles, plastic materials, wires and cables, and heavy metals—which resulted in an interindustry restructuring spree dictated by changing comparative advantages in the domestic economy. Under the promotion schemes of the government, key PC components in particular have become a major area for extensive investment by these traditional firms. Take the Formosa Plastics group, for example. Originating in the petrochemical industry, the Formosa group began to invest in First International Computer in the mid-1980s, which soon became the world's largest motherboard producer and a formidable competitor in the PC assembly industry. It is now extending operations into the realm of 8-inch 16/64Mb DRAM as well as STC LCDs via its main arm, Nan Ya Technology, in anticipation of these sectors' strong growth. As another example, Walsin Lihwa, the second largest cables and wires manufacturer in Taiwan, invested in Winbond (a spin-off from ERSO) in the 1980s and found it one of the biggest cash earners among all its affiliates. Similarly, the Hualon group from the textiles industry recently announced a plan to build an 8-inch wafer fabrication facility, in addition to its established affiliate, Hualon Microelectronics (HMC), which produces semiconductors and is moving fast to charge coupled devices (CCDs). Finally, one of the leading automobile producers in Taiwan, the Yulong group, also bought out Photron Semiconductor recently and is negotiating with OKI to transfer 4Mb and 16Mb DRAM technology from Japan. Even further upstream, China Steel has teamed up with MEMC to set up a joint venture (by the name of Tysil Electronic Materials) for the manufacture of silicon wafers, the first such establishment in Taiwan. Following its lead, one of the world's principal manufacturers of silicon wafers, Komatsu, also moved into Taiwan recently in a joint venture with Nan Ya Technology and the Asia Development Holding Company (to which the Formosa group has a strong tie) to produce silicon wafers directly in Taiwan. With the rich spectrum of vertical linkages thus created, Taiwan's IC and PC industries are now in a position to operate more efficiently than ever before—except for the two still-missing segments of semiconductor manufacturing equipment (capital goods) and systems design (CPUs).

Taiwan's PC Industry at the Crossroads

Taiwan has come a long way in developing its PC and IC industries, but it has not gone far enough. Downstream PC producers still

Figure 6-3. *World Market Shares for Taiwan's Domestic Producers of PC Key Components, 1993, 1995*

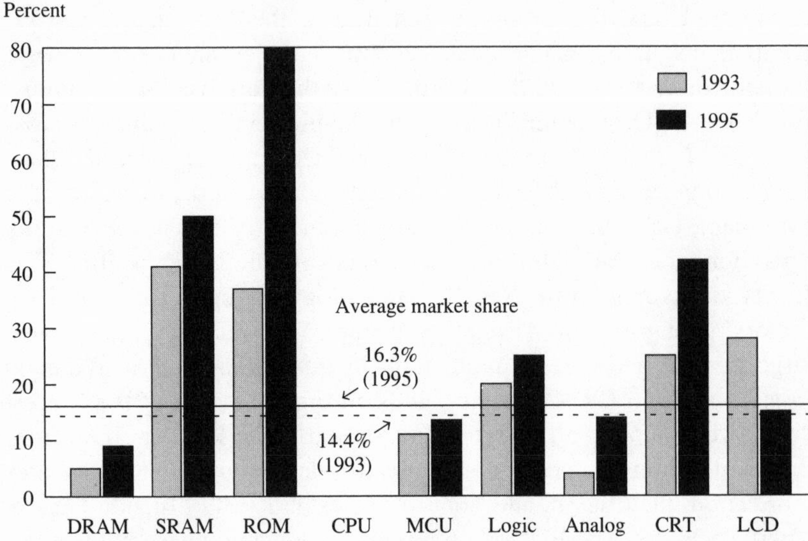

Sources: Electronics Research Service Organization, Taiwan; and Industrial Technology Research Organization, Taiwan.

imported 84 percent of the required information ICs, 85 percent of the LCDs, and 99 percent of the CPUs in 1995. Domestic manufacturers' market share in DRAMs was a mere 8.5 percent in 1995 and close to nothing for CPUs (figure 6-3). The five largest DRAM suppliers in Taiwan are Goldstar, Mitsubishi, NEC, Samsung, and Texas Instruments. The only domestic firms that have achieved some market share are Mosel-Vitelic, Venguard, and TI-Acer. However, in view of the major domestic investment plans in the 16/64Mb DRAM area listed earlier and the fact that Taiwan has managed to keep pace with state-of-the-art process technology in the field (table 6-13),[33] its dependence on Korean and Japanese suppliers for the procurement of DRAMs may fall steadily in the next few years.

The prospect for CPUs is less certain. Currently, U.S. firms such as Intel, Advanced Micro Devices, Cyrix, Motorola, and Texas Instruments are the prime suppliers in the domestic CPU market. These firms occupied more than 99 percent of the Taiwanese market in 1995. The

Table 6-13. *Taiwan's Catch-Up in IC Process Technology, Selected Years, 1976–2000*

	Year when achieved		
Micron	*Taiwan*	*World leader*	*Gap in years*
7μ	1976	...	...
5μ	1980	...	...
3.5μ	1983	...	...
1.5μ	1986	1980	6
1.2μ	1988	1983	5
1μ	1990	1986	4
0.8μ	1991	1990	1
0.5μ	1994	1992–93	1–2
0.35μ	1996/10Q	1996/1Q	.75
0.18μ	2000[a]	2000[a]	...

Sources: Adapted from Electronics Research Service Organization and Industrial Technology Information Services.

a. Expected date for mass production.

main problem here lies not with technology per se but with marketing barriers and intellectual property rights (IPRs). Taiwanese firms proved able to develop independently the world's second prototype of 32-bit PC—after Compaq but ahead of IBM—in 1986.[34] They have also acquired a proficiency in the development of the Chinese language software, which has led many MNCs to use Taiwanese software designers to develop software in both Chinese and other languages not using the Roman alphabet (for instance, Thai and Korean for word processing systems).[35] But Taiwanese producers, as latecomers, simply cannot help stepping on the toes of U.S. firms.[36]

Leading U.S. hardware producers, such as IBM and Intel, and system software makers, such as Microsoft, have long set the global standards for personal computers. This poses a problem of "compatibility" for other firms. On the one hand, the existence of de facto standards in personal computers means that hardware and software producers around the world have already invested large sums in application products designed around those chips, making the costs of conversion to another microprocessor design extremely high (the disadvantage of the latecomer).[37] Moreover, since software development is characterized by large up-front investment but negligible marginal costs of production once the product has been developed, the companies whose operating systems have been adopted as industry standards have a sizable cost advantage. They can spread their development costs over a large number of units, thereby raising the hurdles for new entrants by way of price competition.[38] On the other hand, if a newcomer tries with minor modifications to produce CPUs within the existing framework of design, it then has to face a complicated network of IPRs spawned by the same market leaders. The established producers can easily forestall competition by pursuing prohibitive royalty fees or outright prohibition of usage rights. To circumvent the web of IPRs spawned by these producers, the R&D effort of a new entrant must exceed those of Intel and Microsoft by many times just to break through the existing framework, which again places the new entrant at a cost disadvantage. Market leaders can further block competitors with their well-orchestrated distribution networks and established brand names. Consequently, for a newcomer such as Taiwan, if domestic R&D capacity cannot ensure follow-up development of an entire product series at extremely low costs, Taiwanese CPU producers will not be able to gain trust and loyalty from downstream users at home and abroad and will thus lose the momentum necessary to grow in the long run.

As a result, Taiwan's PC industry is at a fork in the road. Having established itself as a credible manufacturer of PC hardware, it now faces the difficult choice of continuing exploiting its manufacturing capabilities—but increasingly at another geographic location—or making an entry into new product areas characterized by high risk and uncertainty. A comparatively weak marketing ability and limited owner-specific technology are among the main constraints facing Taiwanese firms, and both are directly related to firm size, which, in turn, is endogenous to Taiwan's industry structure.[39] If Taiwan does not upgrade quickly enough, it may come under increasing danger of losing competitive advantage even in the area of hardware manufacturing, because most of its production know-how and managerial proficiency are bound to be transferred offshore to ensure the success of its overseas FDI operations. At the same time the opportunity for product diversification and upgrading seems brighter than ever for Taiwanese firms. The global PC market is being reshaped by buoyant advances in technology, new ideas, and an ever-heightening demand for user-friendly innovations. Recent developments in digital technology and the widespread use of the Internet, in particular, have brought about a rapid convergence between 3-C (computer, communications, and consumer electronics) products, redirecting the global PC market toward a multimedia and network-centered age. Under the new platform, which is only taking shape gradually and leaving many producers confused about its eventual configurations, new products and devices are introduced to the market at a speed baffling even to the most sophisticated suppliers.

The current global structure of production is entering a new phase of change. The United States, having lost its market dominance to its competitors (particularly those from Japan), first in hardware manufacturing and then in key PC components, is now clinging to its leading position in systems software and special chip design to sustain a global market share. While Japan has taken over most of the key component markets, it is increasingly losing comparative advantage in PC hardware manufacturing (figure 6-4). Lower down the value chain, Taiwan (and to some extent South Korea and Singapore) is now assuming the role of a prominent hardware manufacturer and ODM and OEM subcontractor capable of replenishing new peripherals and devices as the market progresses. At the bottom of the global production network stand China, the ASEAN countries, and a host of other developing countries that constitute the latest target for production redeployment by first-tier, second-tier, and third-tier PC firms. The objects of such transplants now consist of standardized peripheral equipment, low-end PC hardware, and some application software designs, but increasingly more advanced models may be relocated to these low-wage countries as market competition becomes more severe.

Within a reasonable time frame, the U.S. and Japan are expected to maintain a controlling edge in global competition in 3-C products. The United States, in particular, is backed further by a winning edge in telecommunications technologies, whereas Japan is likely to remain dominant in consumer electronics. While the two countries vie for leadership in the emerging struggle between multimedia versus network-based PCs, the next tier of countries, including Taiwan and other NIEs ready for the game, stand to gain OEM and ODM opportunities coming from either side and created by whatever market niches open to their manufacturers. At the same time it also seems opportune for these "intermediate" countries to attempt a strong entry into the key components and system software segments of the industry, both as a countervailing act to break Japan's (and the United States') strong hold and as a preemptive move in anticipation of emerging market opportunities. It is only through such efforts that Taiwan, as a latecomer, may have

Figure 6-4. *Present and Possible Future Division of Labor in Global PC Market*

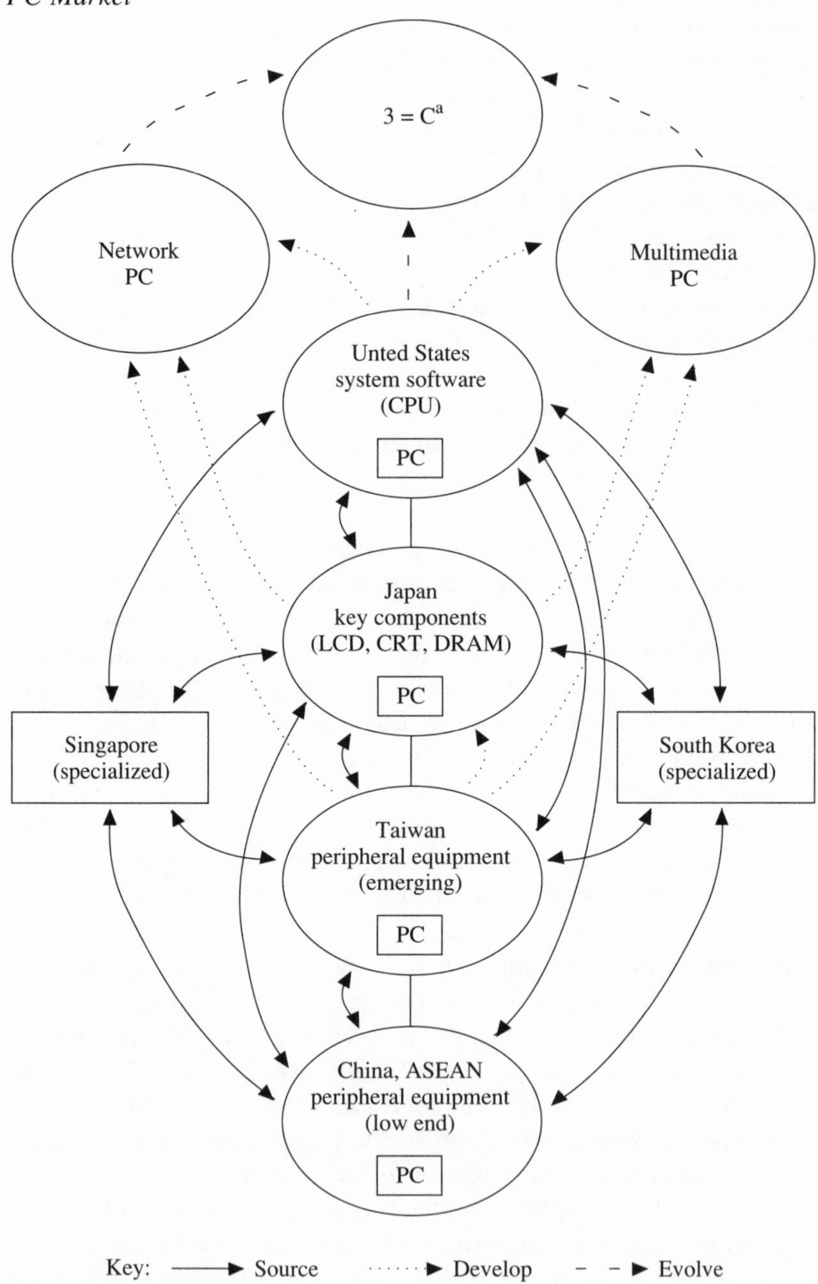

a. Computer, communications, and consumer electronics.

a chance to surpass its present role as a pure "manufacturer" of PC products.

In a slightly different context, Soete, among others, has argued that developing countries with adequate industrial infrastructure and skill levels may benefit from the window of opportunity provided by a new technological paradigm, especially at an early stage of diffusion when barriers to entry are relatively low and markets are in a state of upheaval.[40] The degree of competition that exists in the PC market today has rarely been more intense, and the pace with which ideas and products evolve has never been more acute. In other words, the possibility for making inroads into more advanced areas of production has rarely been more favorable for a country that is prepared to do so. Indeed, with a proven history of indigenous innovative capacity (though often checked prematurely by competitors because of a lack of financial and distributional muscle), Taiwan stands in a strategic position to benefit from its quick time to market, versatile design capability, strong production engineering, and exquisite product targeting, all supported by a highly functional system of institutional infrastructure.[41] To capitalize on that position, however, Taiwan has to overcome its two main ills; that is, it must increase its R&D and put a much greater emphasis on marketing, including establishing its own distributional networks. Acer, for example, has recently teamed up with Legend Computer, the biggest distributor-manufacturer of PC-related products in China, in an effort to preempt Japanese and U.S. distribution networks in the vast Chinese market. Strategies like that, coupled with proper product targeting, may prove to be more important than the simple accumulation of technological capabilities if Taiwan is to compete in the future with the world's most advanced PC producers.

China's Role: Partner or Competitor?

An obvious question arises: if Taiwan can make a quantum leap into the realm of key PC components and product and process innovations, why cannot China do the same thing—and thus undercut Taiwan—given that FDIs are now rushing into the country at a speed unprecedented in any other developing country in the world? Following Motorola, Philips, IBM, Apple, and Hewlett-Packard, among still others, Intel Technology has recently announced a $99 million static random-access memory (SRAM) project in Shanghai, which will complement its current operation in Manila, for the manufacturing of memory chips for PCs, cellular phones, and other digital consumer electronic products. According to Intel vice president Parker, the project will be "no less spectacular than any other high-tech center around the globe."[42] Once Intel is in China, before long most other competitors that have not already done so will be forced to follow suit, as will IC companies from Taiwan.[43] The Chinese policy of "technology in exchange for market," which targets the world's largest electronics multinationals, is likely to reinforce the tendency for such high-tech MNCs to invest and manufacture in China. As the Chinese market develops in both technical sophistication and size over the next decades, will the mantle of electronics leadership eventually pass from the United States and Japan to China?

It must be remembered that the potential for technological leapfrogging in a highly dynamic sector such as the PC or IC industry is particularly subject to the absorptive capacity of the country in question. The country must possess an adequate local base in technological, institutional, and human resources so as to successfully transform foreign example into indigenous industrial activities. Several points may be noted to highlight China's present deficiency to foster such technological and industrial leapfrogging.

First among these may be, ironically, the extent of foreign participation and dominance in the current Chinese electronics industry. MNCs come to China to penetrate its vast internal market, not to disseminate core technologies. Massive amounts of FDI may be pouring in, but research and development activities as well as basic chip design are likely to be retained in the home bases of the investing firms, as exemplified by the experience of previous FDI recipients in the field, including Taiwan, South Korea, and Singapore.[44] Thus in the foreseeable future China will most likely be assigned the role of a regional manufacturer, servicing the Chinese market in particular and under strict control of foreign MNCs. Only gradually will it take on more locally oriented OEM and ODM contracts as its infrastructural and technological capabilities build up. At the same time, however, FDIs could crowd out domestic activities, given the FDIs' competitive strengths in the marketplace and the current supply conditions in China. Singapore provides a good example here. The semiconductor industry in Singapore is represented overwhelmingly by foreign MNCs. Its domestic capacity remained underdeveloped largely because of a brain-drain effect, with foreign MNCs absorbing and preempting the local talent necessary for an indigenous IC sector to thrive.[45] The Singapore experience illustrates the importance of a locally based approach to high-tech industrial development, along with proper human capital accumulation and industrial learning, for subsequent breakthroughs to be possible. At present, the Chinese economy is still plagued by a large state-owned sector in which institutional and structural rigidities abound.[46] The defects of this sector for fostering the growth of a high-tech industry include a sluggish information gathering and assessment mechanism and therefore slow decisionmaking, a lack of efficiency and creativity in general corporate management, and, in particular, the low morale of the work force due to a noncompetitive salary scale. As is generally recognized, a pool of highly skilled, richly experienced, and well-rewarded personnel from home and overseas is one of the most important factors for the rapid development of indigenous high-tech industries.[47] Given its current reward system, however, the state-owned sector can scarcely attract even local talent, let alone foreign-trained expertise, away from MNC entrants into the Chinese economy.

Second, and closely related to the first point, China seems to have placed too much emphasis on generating large industrial conglomerates (implicitly to counter the dominance of foreign multinationals) to the neglect of spontaneous small and medium-size enterprises. By pursuing this Korean (or Japanese) model of industrial development, China may be inadvertently placing itself in a deadlock situation by being forced to choose between two evils: a much bolder embrace of the MNCs (with joint ventures being the norm to stimulate growth in state-owned conglomerates) or a bleak retreat to its own noncompetitive industrial establishments (when an independent development strategy is pursued). The difficulties for present-day China in adopting the Korean model stem, of course, from the fact that its state-owned sector is not as market tested and ready as the Korean *chaebols* in the late 1970s and early 1980s. An alternative, and highly viable, approach to developing high-tech industries is the one provided by Taiwan, which essentially encouraged a pool of innovative and dynamic SMEs to work side by side with large-scale operators such as Taiwan Semiconductor Manufacturing Corporation and United Microelectronics in the semiconductor industry (which were spin-offs from the public research institution, ERSO), and Acer and FIC in the PC industry (which are natural outgrowths of SMEs). This more balanced approach to development (one that is more market oriented and, perhaps, like the U.S.

model) has helped create an environment in which different segments of the PC and IC industries are given the ability and incentives to thrive, each depending on the success of the others. At the same time this approach helped spawn an efficient indigenous production network that greatly enhanced the stability of the entire industry, giving it more structure to sustain market fluctuations and gain international competitiveness. Viewed in this light, the current Chinese policy of "implicit oppression" of private SMEs can prove to be severely counterproductive, if not devastating.

Finally, the erection of high tariff walls and other forms of industrial protection may also work to the detriment of the indigenous growth of the PC industry. Take again the case of Taiwan. It is striking how the PC and IC industries have developed in Taiwan almost without any protective barriers being erected at any time.[48] Such measures are purposely avoided in order to give downstream producers an edge in international competition. As a result, Taiwanese PC producers and chip makers are forced to control production and managerial costs in such a way as to emulate the best practice exercised by some of the top-ranking MNCs.[49] By contrast, protective measures in China have been used extensively both to attract foreign investment and to foster domestic industrial growth. This policy orientation may prove counterproductive, because it could hamper necessary competition from foreign firms and breed inefficiency in domestic suppliers, so that the latter would never become internationally competitive manufacturers. An FDI enclave, rather than a flourishing domestic supply base, might be the result of this policy package of high protection, FDI attraction, and a lopsided reliance on rigid state-owned sectors.

For indigenous capacity to grow in China, the more important factors are the provision of flexible and efficient institutional support, good management, calculated product choice, and strategic corporate planning (and, of course, always with a heavy dose of learning and self-directed R&D). These depend on the overall progress of economic reform in China. Without them—and it may take a long time—it is unlikely that China will become a strong power in electronics except for its manufacturing potential and its role as a regional production base dominated by foreign MNCs. In the interim, if Taiwan is able to combine domestic upgrading and a carefully balanced (and policy-regulated, if necessary) outflow of FDI, China is likely to remain a help rather than a hindrance to Taiwan's future expansion of the PC industry.

Conclusion

The emergence of China and other Asian developing economies in the 1980s has caused a drastic shift in comparative advantages in production and trade for all the regional economies involved. Taiwanese FDI in China has so far been limited to a relocation of lower value-adding production activities, but the process may increasingly demand more substantive transfers of production technology, managerial know-how, and marketing skills. This has put tremendous pressure on investing firms to engage in extensive product realignment at home. The process is reminiscent of Paul Krugman's model of North-South technology trade, in which new industries have to emerge constantly in the North to offset declining comparative advantage in the traditional sectors that are being driven out of existence in the face of low-wage competition from the South.[50] The "monopoly" power of the North is being continually eroded by "technological borrowing" from the South and thus must be maintained by the continual introduction of new products.

In more complicated real-life situations, Taiwan is like a country in the middle, which has its technologies borrowed by the South but which may encounter difficulties borrowing technology from the North. It is my view that for countries in the middle in particular, which do not yet have a proven ability in technology innovation, there is clearly a normative scope for policy intervention. For this reason the emerging pattern of division of labor, though initially induced by free-market forces and played out by independent private actors, may in due time invite public-sector interventions of various sorts.[51] The case of Taiwan versus China—with its peculiar tone caused by close cultural bonds and treacherous political undercurrents—only presents an extreme example of a general phenomenon.

The PC industry provides a small-scale synopsis of the situation facing Taiwan today. Taiwanese small-scale producers often found relocating to China an easier option than upgrading product lines at home. Only the medium-size and large firms showed a tendency to engage in product differentiation or diversification. Under proper policy guidance, though, much interindustry restructuring by large, traditional firms—those coming from textiles, petrochemicals, cables and wires, and heavy metals—made impressive advances in the PC and IC industries. The investment booms thus created in those sectors accounted for almost one-third of Taiwan's private industrial investment in the past few years and have rendered the PC and IC industries the most dynamic sectors in the economy. Important areas for investment include large-size CRTs, LCDs, LEDs (light emitting diodes), and DRAMs, as well as silicon wafer plants. All these seem suited to Taiwan's current and future comparative advantage. Thanks also to buoyant advances in technology and emerging market niches, Taiwan is in a favorable position to benefit from its quick time to market, versatile design capability, manufacturing expertise, and ingenious institutional innovations to develop new competitive edges in the upcoming 3-C era. China, in this context, can provide Taiwanese firms with an excellent location for intraregional division of labor, given its abundant labor supply, a well-trained reservoir of local talent, a vast internal market, and an additional advantage of minimal (implicit) entry costs.

At the same time the precariousness of the future competition facing Taiwan can hardly be exaggerated. Currently Taiwan lags considerably behind its potential competitors in many areas: technology, R&D, marketing, and general corporate strengths. The IC industry in Taiwan has recently set the target of reaching an output value of $25 billion and a world market share of 8–10 percent by the year 2000, greatly advancing from its 1995 status of $3.5 billion and 2.8 percent. The PC industry has similarly set a target of $65 billion and a world market share of 14 percent by 2000, up from its 1995 status of $21 billion and 7 percent. Since the PC and IC industries now constitute the two most lucrative and dynamic sectors in Taiwan's economy, whether they can achieve their goals will have an important impact on the total performance of Taiwan's manufacturing sector. As for China, its role in the emerging pattern of the intraregional and interregional division of labor is likely to remain that of a manufacturing base—albeit a gigantic one—until it is able to rid itself of the institutional and structural rigidities embedded, in particular, in its state-owned sectors. As a related point, though it is true that in Taiwan a new emphasis on large-scale operators (and operations) may be forthcoming in its industrial development strategy, it is equally true that in China what is missing for a successful fostering of an indigenous PC supply base is a growing pool of spontaneous and versatile SMEs. Shifting comparative advantage across geographic regions not only engenders a new pattern of division of labor among different locations but also implies a need for fundamental changes in the organizational structure of the economies involved. In this respect, too, both Taiwan and China have much to accomplish in the years to come.

Notes

1. In his eclectic theory Dunning identified three necessary conditions for an act of FDI: an *ownership* advantage of the FDI firm, a *location* advantage of the FDI host, and an *internalization* advantage of the FDI act. See John Dunning, "Trade, Location of Economic Activity, and the MNE: A Search for an Eclectic Approach," in Berlin Olin Hesselborn and Per Magnus Wijkman, eds., *The International Allocation of Economic Activity* (Holms and Meier, 1977).

2. Stephen Hymer, *The International Operation of National Firms: A Study of Direct Foreign Investment* (MIT Press, 1976).

3. For a detailed account of the investment behavior of Taiwanese SMEs in China, including their motivations, limitations, and trade consequences, see Chin Chung, "Double-Edged Trade Effects of Direct Foreign Investment and Firm-Specific Assets: Evidence from the Chinese Trio," paper presented at the International Conference on Sino-U.S. Economic Relations, sponsored by the Center for Asian Pacific Studies, Lingnan College, Hong Kong, June 21–23, 1995.

4. This was compared with the Japanese situation in the late 1960s and early 1970s, when SMEs constituted the norm of Japanese investment. I argued that for Japan an equally exogenous strand of advantage was injected from outside the FDI carriers in the form of *keiretsu* assistance or government guidance, or both—in contrast to advantage from a cultural and linguistic proximity for Taiwanese SMEs operating in China. See Chung, "Double-Edged Trade Effects."

5. The figures are obtained from an official survey of all FDI operations in China as of February 1993. See Charng Kao and Shih-Ing Wu, *An Investigation of Foreign Direct Investment in Mainland China* (in Chinese), project report commissioned by the Investment Commission, Ministry of Economic Affairs (Republic of China, 1994).

6. See Chin Chung, "Taiwan's DFI in Mainland China: Impact on the Domestic and Host Economies," paper presented at the third annual conference of the Chinese Economic Association, London, December 1991; later collected in Thomas Lyons and Victor Nee, eds., *The Economic Transformation of South China: Reform and Development in the Post-Mao Era* (Ithaca: Cornell East Asian Series, 1995), pp. 215–42.

7. For details of the setup of this model and a discussion of the underlying assumptions, see Chung, "Taiwan's DFI in Mainland China," pp. 226–30.

8. All things considered, the net effect on home production is given by the equation $\Delta Y = D\Delta T - D[\Delta K_i(Y_i/K_i)]$, where Y_i/K_i is the domestic output-capital ratio (which may or may not be the same as $(Y_i/K_i)^*$. The first part of this equation shows the increase in home production induced by an increased demand for its exports (in the form of intermediate inputs) by FDI operations overseas, while the second part shows the decline in home production due to a relocation of its capital stock and production activity. For a full derivation of this equation, see Chin Chung, "Macroeconomic Impacts on the Domestic Economy of Taiwanese DFI towards Mainland China" (in Chinese), in Chiu Lee-in Chen, ed., *Taiwan's Direct Investment in Mainland China: Policy and Strategies* (Taipei: Chung-Hua Institution for Economic Research [CIER], 1991), pp. 91–120.

9. Operationally, the d_is enter the equation in the form of a 29 × 29 diagonal matrix with the d_is sitting on the diagonal and zeros everywhere else.

10. The coefficients used in the estimation of this model are derived from interviews and surveys conducted by the Chung-Hua Institution over the past five years. See, for

example, Tsong-ta Yen, Y. J. Lin, and C. Chung, *A Study of Taiwanese Investment and Trade Relations with Mainland China* (in Chinese) (Taipei: CIER, 1992); and C. H. Ouyang and others, *A Trade-Warning System for Monitoring Taiwan-Mainland Economic Interdependence and Its Applications* (in Chinese) (Taipei: CIER, 1990).

11. By using primarily a survey method, Kao, Lee, and Lin reached broadly similar estimates, especially in terms of Y_i^* and X_i^*, to the ones presented here. See C. H. C. Kao, Joseph S. Lee, and C. C. Lin, *An Empirical Study of Taiwanese Firms' "Mainland Investment" on Industrial Upgrading in Taiwan and on the Vertical Division of Labor across the Taiwan Strait,* project report commissioned by the Ministry of Economic Affairs (Republic of China, 1993).

12. Chung, "Taiwan's DFI in Mainland China," pp. 237–38.

13. According to Taiwan's input-output table for 1989, these are all labor-intensive sectors with capital-labor ratios below 0.8.

14. For a detailed discussion of these recent changes, see Chin Chung, "Industry Characteristics and FDI Strategy: A Three-Way Typology of Taiwanese Investment in Mainland China," in Joseph S. Lee, ed., *The Emergence of the South China Growth Triangle* (Taipei: CIER, 1996), pp. 287–326.

15. Before October 1990 FDI on the mainland was considered illegal by the Taiwanese government even though cumulative contractual investment toward China was already approaching $2 billion by the end of 1990. In October 1990 the government legalized 3,353 manufacturing items, mostly labor-intensive products, as permissible FDI objects in China by Taiwanese firms. This "positive list" was also applicable to publicly listed companies, provided their China-bound investment did not exceed 20 percent of their registered capital.

16. Paul A. Samuelson, "International Trade and Equalization of Factor Prices," *Economic Journal,* no. 68 (1948), pp. 163–84.

17. Ministry of Economic Affairs, Department of Statistics, *Monthly Statistics of the Import/Export Trade, Taiwan Area and Statistics of Industrial Production Monthly, Taiwan Area* (Republic of China, September 1996).

18. The role of MNCs in the PC industry has been rather indirect in Taiwan. They affected the local industry mainly through OEM-ODM procurement activities with different degrees of technological assistance. See, for example, Momoko Kawakami "Development of the Small- and Medium-Sized Manufacturers in Taiwan's PC Industry," Discussion Paper 9606 (Taipei: CIER, November 1996).

19. Kawakami, "Development of the Small- and Medium-Sized Manufacturers," pp. 25–27.

20. C. S. Chou, *The Computer Legend of Stan Shih* (in Chinese) (Taipei: Leinking Publishing, 1996), pp. 119–28.

21. There were only a few important foreign-affiliated PC producers in 1994, ranging from the fourth place to the thirty-first in industry ranking, including Digital Equipment International, AOC International, and Logitech Far East.

22. Here is a famous example of this adaptability: while it took about a year and half (from late 1982 to early 1984) for Taiwanese firms to catch up with the world's first model of IBM-compatible 80286 PC XT (introduced by Compaq), it took less than a month for them to develop a compatible Pentium PC (from April to May 1993) .

23. For example, Acer Peripherals, a subsidiary of Acer Inc., specialized in the production of PC peripherals and grew to be the largest local supplier of CD-ROM drives

in 1995—one year after it decided to enter the market—with an annual output of 1 million sets. See Chou, *Computer Legend of Stan Shih,* p. 324.

24. In 1995 the Taiwanese IC industry's total R&D expenses amounted to $230 million, whereas Intel spent $1.3 billion. Data obtained from Industrial Technology Research Institute, Industrial Technology Information Services, Taiwan.

25. This figure came down from some 70 percent in the early 1990s, when many Taiwanese PC producers were vigorously grabbing OEM and ODM opportunities arising from an acute price competition in the global PC market ignited by Compaq.

26. Starting in 1995–96, as a result of fierce market competition, some of the emerging products at home, including drawing interface and CD-ROM drives, also began to move offshore.

27. Officially, Taiwanese PC firms were forbidden by law to conduct FDI on the mainland until as late as 1992. The 1990 positive list of 3,353 items did not include the PC products, which were deemed to be high-tech by the authorities. Still, many Taiwanese PC producers, especially smaller ones, took advantage of the low production cost in Guangdong and Fujian by conducting *lailiaojiagong,* or outward processing, for various low-end PC items.

28. A recent case study by Shu showed that as of 1993 nineteen of the thirty-eight major monitor producers in Taiwan had already established some form of cooperative relationship (including outward processing) with Chinese operators, and another six firms were making plans to invest in China. See E. Shu, "DFI in Mainland China: A Case Study of the Monitor Industry," in Chin Chung and others, *The Industrial Competition across the Taiwan Strait: Past Trends and Future Prospects* (in Chinese), project report commissioned by the Council for Economic Planning and Development (Republic of China, 1994), chap. 8.

29. Taiwan Economic Research Institute, *Almanac of Taiwan's Information and Electronics Industry* (Taipei, 1996), p. 118.

30. Shu, "DFI in Mainland China," pp. 32–40.

31. When asked to compare the investment environment among Taiwan, ASEAN countries, and China, Taiwanese hardware producers typically find China superior with regard not only to labor and land costs and ease in daily communications but also to tax benefits for foreign investors. The only area in which China has scored lower than Malaysia is in the provision of public infrastructure. See Institute for Information Industry, Market Intelligence Center, *Data Bank on the Production and Sales of the Information and Electronics Industry* (in Chinese), project report commissioned by the Ministry of Economic Affairs (Taipei, 1994), p. 15–17.

32. The idea was first proposed in 1991 but was not fully adopted by the government until late 1992. See Chung, "Macroeconomic Impacts on the Domestic Economy"; and Yen, Lin, and Chung, *Study of Taiwanese Investment and Trade Relations.*

33. On the one hand, the government-supported research institution, ERSO, played a pivotal role in accelerating the development and dissemination of the IC processing technologies in Taiwan. On the other hand, tremendous growth in the indigenous IC industry, together with the handsome profits to be earned, also enabled individual firms to accumulate technology by initiating self-directed R&D and by forging strategic alliances with foreign multinationals.

34. Acer launched its first 32-bit microcomputer in 1986 and took all of its Western competitors, including IBM and Compaq, by surprise. This, however, had come about

only naturally, since Acer had a proven history of being the first in launching new products. It developed the first Chinese operating systems for computers in 1980, as well as its own 4-bit, 8-bit, and 16-bit microcomputers in subsequent years. In 1988 Acer further announced the world's first PC86 chip designed for PS/2 model 30 PCs through its central research hub, Acer Laboratories. See Matthew, *High-Technology Industrialization in East Asia,* pp. 90–91. See also "Big Blue's Shake-up," *Economist,* November 30, 1991, p. 19.

35. Dieter Ernst, "Automation and the Worldwide Restructuring of the Electronics Industry: Strategic Implications for Developing Countries," *World Development,* vol. 13 (March 1985), pp. 333–52.

36. For example, United Microelectronics Corp. (UMC), the largest own-brand IC manufacturer-designer in Taiwan, developed its own CPU in 1994 after painstaking endeavor. But the company was forced to stop entry into the industry because of an alleged infringement of Intel's international property right. Similarly, when Acer introduced its first IBM-compatible PC in 1983 and was receiving rising orders from overseas (including from U.S. companies such as NCR), its products were detained by the U.S. customs because of an alleged IPR violation against the basic I/O system adopted by IBM. Even though the system used in Acer PC was developed independently in Taiwan (by ERSO and at a cost to Acer), it was not developed under a "clean room" concept. The dispute was later resolved by Acer's obtaining a different system developed by Data Resources. In both cases, however, evidence of plagiarism was never really established.

37. As O'Connor explained it, since the established industry standards already possess a voluminous library of programs written to run under them, a new entrant would need to convince both the applications software writers and hardware producers of the viability of its own product so as to induce them to develop new applications to run on it well before the product is shipped to market. In other words, only an operating system that represents a true technological breakthrough beyond the state of the art is able to induce such a result. See David C. O'Connor, "The Computer Industry in the Third World: Policy Options and Constraints," *World Development,* vol. 13, no. 3 (1985), pp. 311–32.

38. See O'Connor, "Computer Industry in the Third World," pp. 314–15.

39. Taking (and assigning) OEM orders may be one source of this perpetuated power imbalance. Taiwanese PC producers often fall into the OEM trap, in which market orders are steadily forthcoming and production capacities are being filled, but the firm in question is left with little incentive or free resources to pursue a more independent route of corporate development.

40. Luc Soete, "International Diffusion of Technology, Industrial Development, and Technological Leapfrogging," *World Development,* vol. 13, no. 3 (1985), pp. 409–22.

41. For an excellent discussion of the institutional innovations of the Taiwan IC industry, as well as an in-depth comparison between the different paths taken by Taiwanese and Korean IC firms, see John A. Matthew, *High-Technology Industrialization in East Asia: The Case of the Semiconductor Industry in Taiwan and Korea,* Chung-Hua Institution for Economic Research (CIER) Contemporary Economic Issues Series 4 (Taipei, December 1995). See also Robert Wade, *Governing the Market: Economic Theory and the Role of Government in East Asian Industrialization* (Princeton University Press, 1990).

42. *Commercial Times,* November 14, 1996.

43. So far FDI made by Taiwanese IC firms has been directed mainly toward the advanced-country market—the United States and Europe in particular—in an attempt to gain technology or to forge strategic alliance with local firms. China remains a high-risk region for these firms, owing both to the tremendous stakes (in terms of financial outlays) involved and to the firms' relative inability to spread those risks.

44. As noted elsewhere, the extent to which each of these economies has been exposed to FDI by MNCs is different, as is the resulting impact on the development of indigenous capacities. See Michael Borrus (chapter 5); Matthew, *High-Technology Industrialization in East Asia;* and Dieter Ernst, "What Are the Limits to the Korean Model? Berkeley Roundtable on the International Economy, University of California, Berkeley, 1994.

45. See Shun-chiao Chang, "The Analysis of Taiwan's Semiconductor Industry: A Test of the Technological Leapfrogging Theory," CIER, Taipei, 1996.

46. See chapter 8 for a more detailed discussion of the current electronics industry in China.

47. This is certainly true for Taiwan and Singapore, where a significant backflow of talent contributed to the development of the local PC and semiconductor industries. It is also true for the United States, where many U.S.-trained Asian engineers and programmers worked in the Silicon Valley and greatly increased U.S. competitiveness in the PC and semiconductor industries.

48. Taiwan's tariffs on IC and PC products have traditionally been low, at about 5 percent in the 1980s and averaging 1 percent in 1996.

49. See Borrus's discussion in chapter 5.

50. Paul Krugman, "A Model of Innovation, Technology Transfer, and the World Distribution of Income," *Journal of Political Economy,* vol. 87 (April 1979), pp. 253–66.

51. The Information Technology Agreement recently proposed by the United States in the first ministerial meeting of the World Trade Organization provides a good example of this point.

CHAPTER SEVEN

Partners for the China Circle? The East Asian Production Networks of Japanese Electronics Firms

Dieter Ernst

THE CHINA FEVER that has raged through the Japanese industry over the last few years has drastically changed the locational patterns of Japanese investment within East Asia. The share of China in the investment of Japanese electronics firms abroad has increased by leaps and bounds: from the meager 0.6 percent of 1990 (the year after the Tiananmen massacre), it reached almost 7 percent in 1995, catching up fast with the 7.7 percent share of ASEAN countries.[1]

Japanese electronics firms have thus substantially extended the geographic coverage of their East Asian production activities.[2] At the same time they have proceeded to integrate their erstwhile stand-alone operations in individual host countries into increasingly complex international production networks.[3]

These changes have had important implications. Japanese electronics firms are now in a much better position to act as partners for the China Circle. To cope with increasing complexity, their Asian production networks have become more open and locally embedded. Japanese firms now have a vested interest in developing and harnessing the region's resources and capabilities. They are also now more willing to interact with local companies and to increase their Asian value-added. The Asian production networks of Japanese electronics firms no longer exist in splendid isolation. They now interact with a variety of newly emerging production networks in Asia that have been established by firms from Korea and the China Circle.[4] Originally focused on subcontracting and original equipment manufacturing (OEM) arrangements, such interactions now cover more complex stages in the chain of values, including engineering and product development.

How does this situation match with the widespread perception that, especially in Asia, Japanese firms have kept their production networks as closed as possible to outsiders by centralizing almost all strategic decisionmaking and high value-added activities in Japan? It has been argued that Japanese electronics firms have fallen back relative to their U.S. competitors in large part because they failed to establish an equally efficient regional supply base in Asia, a failure primarily due to the relatively closed and Japan-centered nature of their Asian production networks.[5]

In my view this argument needs to be taken with a grain of salt. This chapter adds two important qualifications. First, a clear-cut distinction between closed Japanese and open U.S. production networks in Asia has existed only for a fairly short period, roughly from 1986 till about 1992. Both before and after this period, Japanese production networks in Asia were fairly open and locally embedded. Second, one should also not underestimate the capacity of Japanese firms for rapid learning and organizational adjustment.[6] They have clearly understood the constraints that result from closed, Japan-centered production networks. Over the last few years they have seriously moved to establish a regional supply base in Asia to improve their access to the region's capabilities and contested growth markets. Japanese electronics firms now have a good chance to recapture terrain they lost during the 1980s.

The chapter is organized as follows. In the first two sections I trace the development of Japanese production networks in Asia through three different periods. Each of these periods reflects fundamentally different strategic rationales for engaging in Asian production activities and hence gives rise to very different incarnations of Japanese production networks. The first section deals with the period of domestic market orientation, which lasted roughly until the early 1980s, and with the shift to export platform production that gathered momentum especially after 1986. The second section deals with recent developments since

1991, the year of the "bursting of the bubble economy" in Japan. In the next section I discuss some possible implications for the China Circle and conclude with a brief discussion of new opportunities and challenges that are likely to emerge for Taiwanese electronics firms.

From Domestic Market–Oriented to Export Platform Production

Over the last three decades Japanese electronics firms have substantially extended their overseas production activities in East Asia. Originally, tariff hopping and attempts to reap the windfall profits available in highly protected domestic markets were the main motivations. Since the late 1970s, however, almost all the leading Japanese electronics firms have begun to invest in export platform production activities, often in a close symbiotic relation with their main small and medium-size Japanese component suppliers. Especially since 1985, periodic yen appreciations have played an important catalytic role and have led to a massive expansion of such export-oriented foreign direct investment (FDI). Japanese production networks started out with a loose, locally embedded structure during the period of domestic-oriented production. That changed dramatically once the focus shifted to export platform production, which led to the establishment of highly centralized governance structures with very limited local roots.[7]

Domestic Market–Oriented Production

There is widespread consensus that once they move abroad, Japanese companies are in general less obsessed with equity control than are U.S. companies.[8] This fact obviously reflects the heavy focus on the domestic markets that has characterized Japanese production in Asia until recently—the penetration of these heavily protected markets frequently requires local partners. A weaker preference for equity control may also reflect important differences in the domestic capital markets of Japan and the United States and the fact that Japanese firms, until quite recently, had ample access to patient capital.

Some observers also claim that Japanese firms, especially in East Asia, are more willing than Western firms, particularly American ones, to engage in joint ventures and other forms of interfirm cooperation;

they are thus presumably more locally embedded than their Western counterparts. Take for instance a widely quoted book on Asian business practices by two researchers from INSEAD. “Accustomed to operating in their home market through an extensive network of cooperation agreements, [Japanese firms] show a higher propensity to enter partnerships in Asia than American firms.”[9] Until about the mid-1980s the electronics industry fits quite well with this general perception of relatively loose and open governance structures. Japanese subsidiaries in Asia had fairly strong local roots, as long as their main objective was the penetration of protected domestic markets in this region. In most cases, joint ventures with local partners were necessary to implement such rent-seeking investments. Local partners provided access to distribution channels and facilitated relations with the government.

Throughout this period Japanese affiliates in Asia had considerable autonomy in making decisions, not only on how to handle employment, work practices, and salary but also on how to organize production, support services (quality control and maintenance), and procurement. Many local linkages were generated by these investments; local content was substantial, giving rise to the development of some domestic support industries, especially for low-end general purpose components. This development often came, however, at the expense of cost efficiency and quality, which, because of the heavy protection provided to such import substitution markets, were only of secondary concern.

Two companies have pioneered in such domestic market–oriented production facilities in Asia: the huge and powerful Matsushita group and Sanyo, a company that, owing to its smaller size, has been the first to develop a regional supply base in East Asia.

In size, Matsushita Electric Industrial Company (MEI) stands out among its Japanese competitors, being surpassed only by Nippon Telegraph and Telephone. With regard to both consolidated sales and employment, MEI is one of the biggest companies in the world electronics industry.[10] For some time this giant conglomerate has been run like a loose network of (almost) independent business units, with headquarters, particularly under the charismatic leadership of Konosuke Matsushita, playing the neutral role of arbiter. As long as markets kept growing, this loose network organization was widely considered to be a great strength, for it enabled the company to remain reasonably flexible despite its huge size. Indeed, until quite recently Matsushita was considered a role model in the Japanese literature on networking strategies; in analogy to Toyota's role in the car industry, the literature singles out Matsushita as the pacesetter for the electronics industry.[11] But once demand growth could no longer be taken for granted, the debate fundamentally changed. It now focuses on the hidden costs of excessive decentralization and decision autonomy, such as duplication, forgone economies of scale, and self-generated price pressures.

Matsushita's core competencies are size-related advantages in distribution and high-end manufacturing (in specialized affiliates like Matsushita Kotobuki). Because of its tight control over the domestic distribution channels for consumer electronics, Matsushita used to be under much less pressure to become a low-cost leader. That explains why its expansion into East Asia was for a long time geared primarily to the heavily protected domestic markets rather than to the establishment of low-cost export platform production.

Matsushita's involvement in East Asia started in the early 1960s, with minority joint ventures strictly targeted at the heavily protected domestic markets.[12] The so-called mini-Matsus originally produced simple products like batteries, radios, electric fans, rice cookers and other low-end home appliances, small television sets, and some related components. The real breakthrough came in September 1965, when Matsushita established its first large manufacturing plant, MELCOM, in Malaysia.[13] Most earlier plants had been restricted to "screwdriver" assembly with limited local linkages.[14] MELCOM established a radically different pattern that since then has characterized domestic market–oriented Japanese affiliates in Asia. Most of the key components were now produced in-house—that is, by the overseas affiliate—and involved metal stamping and forming, surface treatment (electroplating and painting), plastic injection, and die casting. Over time many of these activities have been outsourced to local suppliers, giving rise to a large body of domestic support industries. In addition to procurement, decisionmaking autonomy was also gradually increased for marketing and distribution and for local investment. Already in June 1974 Matsushita had established its regional headquarters in Singapore, well before any of its main competitors followed suit. Right from the beginning, these headquarters were endowed with fairly broad responsibilities, covering procurement, inventory control for parts and components, technical guidance and training for maintenance, quality control, and support for negotiations with local authorities.[15]

Sanyo went even further than Matsushita in developing early on a locally embedded Asian production network.[16] Sanyo's founder was the brother-in-law of Konosuke Matsushita, who had taken over most of the operations of the Matsushita group when it was broken up under the U.S. Occupation Authority. In Japan, Sanyo always remained a second-tier competitor. One way for it to compensate was to make an aggressive shift into international production. Sanyo's move to East Asia preceded that of Matsushita. It also started much earlier with fully integrated local assembly. Already in the early 1970s Sanyo developed the so-called one-third strategy for manufacturing capacity: one-third domestic manufacture for the domestic market, one-third domestic manufacture for foreign markets (especially higher-end segments), and one-third foreign manufacture for foreign markets. What distinguishes Sanyo from Matsushita is the higher degree of vertical integration that is typical for Sanyo's overseas affiliates. During the second half of the 1970s Sanyo began to develop a regional component supply base in East Asia, much earlier than any of its Japanese rivals.

A Belated Shift to Export Platform Production

After 1985 rent seeking gave way to a radically different concern. Once the focus shifted to export platform production, Japanese firms insisted on 100 percent affiliates or at least majority joint ventures. Governance structures became highly centralized, and outsiders had limited opportunities to enter these production networks. For a short time, roughly from 1986 till about 1991, there was a distinct shift in the organization of Asian production networks. Japanese electronics firms chose a centralized approach with limited roots within the region, while American companies relied more on decentralization, which gave rise to a certain amount of local embeddedness. There are two reasons why Japanese electronics firms choose to rely on centralized control: the vintage factor and proximity.

The Vintage Factor: Latecomers in International Production. What distinguishes Japanese electronics firms is their belated shift from exports to international production. Until the mid-1980s Japanese electronics firms stubbornly resisted the move to international production, trying to reap maximum advantage from basic features of their domestic production system that they knew would be difficult to reproduce abroad.[17] Once this shift finally occurred, Japanese firms were under tremendous pressure and did not have time to proceed in a gradual manner. Under the impact of the yen appreciation,[18] Japanese producers of consumer electronics were quickly losing market share in the United States and Europe, especially to the aggressive new competitors from South Korea and Europe. A quick response on a massive scale was required to roll back these challengers. The solution was to establish huge export platform plants in lower-cost production sites in Southeast Asia that would drastically cut production costs and thus help to sustain market share in the United States and Europe. Production expansion had to occur quickly, and cost and quality had to be tightly controlled. Under such conditions, tight centralized management control exercised by the parent company was a rational choice.

We know from innovation theory that firms need time to develop their capabilities.[19] Time is of even greater importance for developing a firm's capacity to manage international production; hence the importance of the vintage factor. John Stopford argues that "firms progress over time from the simplest to more complex forms [of international production networks] as they learn how to manage [them]." Such learning also takes place in the foreign affiliates: "As skills and resources accumulate within the various foreign units, new options and more complex projects can be undertaken without relying heavily on the parent organization for help and guidance."[20] The result is that latecomers to international production are likely to differ in their organizational approaches from firms that have had a much longer learning experience.[21]

Japanese electronics firms obviously did not have time to follow the gradual approach described in the textbooks on foreign investment. Their first response was to try to transplant key features of the Japanese production system with as little change as possible. They soon found out that this does not work. The main aspects of the institutions in Asian host economies, like labor and capital markets, were simply too different to sustain the wholesale transfer of the Japanese production system. Japanese firms have thus been forced to adjust many of the original features of their domestic organization. The result has been the spread of hybrid forms of organization that require especially strong forms of control.

Proximity. A second factor that explains the closed and Japan-centered nature of Japanese production networks in Asia is proximity, which has facilitated centralized control. The scope for centralized control diminishes with increasing distance. Once a firm extends its value chain across national boundaries, it is faced with complex coordination problems and the risk of abrupt disruptions.

Four sources of disruption are apparent: those caused by suppliers, either through late delivery or through the delivery of defective materials; unforeseen fluctuations in demand and abrupt changes in demand patterns; various production problems that result from the transfer of immature products and production processes; and abrupt changes in management decisions, such as last-minute corrections of product launch dates and performance features. Firms have tried to reduce the likelihood of such disruptions, but so far with only limited success.[22] Although production-related disruptions decline with increasing product maturity, that has not been true for demand-related disruptions and for abrupt changes in management decisions that have been imposed by financial markets.

Japanese firms are in a much better position to manage these risks than are American and European firms. Japanese firms can control their East Asian affiliates from Tokyo, because the region is part of the same time zone: as a rule of thumb, a Japanese parent company is willing to loosen and decentralize control only once the affiliate is more than six hours' flying time from Tokyo. Both in Asia and in Europe U.S. firms never had this option. Probably that is one of the main reasons why, early on, companies like Intel and Motorola were willing to grant some decisionmaking autonomy to their Asian affiliates.[23]

Shortcomings of Centralized Control

Throughout this period Japanese electronics firms all tried to keep their Asian production networks as closed as possible to outsiders by centralizing almost all strategic decisionmaking and high value-added activities in Japan. As a result, local content remained limited, and most components and materials were imported from Japan. One indicator is that Japanese electronics affiliates in Asia relied much more on component imports from Japan than did similar affiliates in North America and Europe. While the share of components in Japan's electronics exports to Asia has exceeded 30 percent, it has been less than 15 percent for Japan's electronics exports to North America and Europe.[24] Another indicator is that component exports were by far the main cause for Japan's huge trade surplus in electronics with Asia. In

1990 components were responsible for 55 percent of Japan's trade surplus in electronics with Taiwan, and for almost two-thirds with Malaysia.[25]

In their Asian export platform affiliates, Japanese electronics firms usually relied much less on local managers and engineers than did their American and European counterparts. The Japanese tightly controlled their Asian affiliates, leaving them little scope for autonomous decisions; the transfer of technological capabilities remained limited and hardly went beyond on-the-job training and basic manufacturing support services.[26]

The same has also been true for some of the organizational innovations that have been the hallmark of the Japanese production system. In most Asian affiliates of Japanese electronics firms, there was no attempt to establish seniority-based wage systems, job rotation, and life-long employment. Surprisingly, that has also been true for quality-control circles and just-in-time management techniques. Often a crude Fordism prevailed, at least during the initial phase.[27] This contrasts with the situation in the United States and Europe, where Japanese firms have made serious attempts to transfer key elements of their domestic production system and to adapt them with great care to the peculiarities of local institutions, policy frameworks, and labor markets.[28]

As long as the focus was on export platform production, therefore, Japanese electronics firms tried to minimize the transfer of activities in the value chain to East Asia. Yet doing so came at a heavy cost. It prevented Japanese firms from harnessing the resources and capabilities that had been accumulated in East Asia; it slowed down the penetration of the increasingly important growth markets of the region; and it obstructed attempts to establish a regional supply base and to improve the specialization of these Asian production networks.

Recent Changes: Increasing Specialization and Local Embeddedness

Since 1991 the combined effect of the yen appreciation, the bursting of the bubble economy, and the domestic recession has transformed Japan's outlook. Japanese firms have been forced to change key features of their Asian production network and have shifted to more open and locally embedded networks. Leading Japanese electronics companies have apparently made a strategic decision that from now on Asia will play a critical role. Lower labor costs continue to matter. Of equal importance, however, are the following four objectives: a regionalization of procurement, the penetration of Asia's contested growth markets, attempts to harness the region's improved capabilities, and a shift to more decentralized governance structures. Coping with these conflicting requirements demands a greater focus on a regional strategy.[29]

Catalytic Effect of the Yen Appreciation: The Regionalization of Procurement

The most immediate concern is the reorganization of procurement. The yen appreciation has acted as a powerful catalyst: it eroded the cost competitiveness of the Japanese production base, and it drastically increased the price of components and capital equipment that Japanese affiliates in Asia imported from Japan. This has led to four important changes in procurement patterns: the parent companies in Japan have increased their imports from Asia, both for final products and components; the major corporations are developing more systematic regional procurement strategies; Japanese component suppliers are redeploying production to Asia, with some of them beginning to develop their own regional production networks, often in close interaction with local producers; and Japanese affiliates in Asia have substituted procurement from local sources for some of their component imports from Japan. That has set in motion a complex process of regionalization of procurement.

Increasing Imports from Asia. Until recently Japan imported only a tiny fraction of the electronics goods and services it consumes at home. This is in sharp contrast to the high-import propensity that characterizes that electronics sectors of the United States and Europe. Asia, despite its close proximity to Japan, played a much less prominent role as a source of electronics imports for Japan than it did for the United States. Since 1992, though, Japan's imports of electronics products have grown rapidly, primarily from Asia. Japan's import ratio for electronics increased from 10.3 percent in fiscal 1985 to 17.5 percent in 1993.[30] The most rapid increase occurred in components, for which the ratio went from 16 percent to over 35 percent. Most of these components are lower-end, general purpose electronic components from East

Asia. The import ratio also increased for consumer electronics and household appliances: from 2.1 percent in 1985, it rose to 10.2 percent in 1993.[31] The share of imports from Asia surged from less than 31 percent in 1988 to more than 44 percent in 1993.[32]

Asia has also been able to strengthen its role as a supply base for components and PC-related products. While in 1988 the United States was the only source of imported integrated circuits and computers, today Japan imports roughly the same number of ICs and computers from Asia and from the United States. Korean chip makers have especially benefited, accounting for almost 28 percent of total chip imports (mainly DRAMs) in 1995.[33] The share of higher-end and technologically more complex products in Japanese imports from East Asia has increased considerably. Take the case of Fujitsu, which from 1994 to 1996 was able to triple its share of the domestic PC market in Japan.[34] Besides the innovative design features of its product, the main reason for Fujitsu's success was arguably a radical shift in component procurement. While in 1994 almost all PC-related parts and components were sourced in Japan, this dramatically changed two years later; in the first quarter of 1996, Fujitsu estimated that 95 percent of the parts for the PCs sold in Japan were imported, most of them from Taiwan.[35]

Regional Procurement Strategies. A questionnaire survey conducted by the Japan External Trade Organization (JETRO) in October 1993 found that more than 80 percent of Japanese firms with substantial overseas production activities indicated that during the next five years they would considerably increase their current levels of international procurement.[36] About one-quarter of the responding companies claimed they were planning to double their current levels of international procurement. This claim is confirmed by the 1996 annual questionnaire survey of the Export-Import Bank of Japan. The survey states, "It is noteworthy, in particular, that reinforcing parts procurement bases overseas is the most important strategy for the automobile assembling industry and the electric/electronic industry." And it concludes that today "Japanese companies have tried to reconstruct their advantages internationally through building more flexible and open international networks between parts suppliers and assemblers."[37]

The challenge is how to achieve these cost savings without losing too much in terms of quality, speed, and reliability of delivery. This is a formidable challenge. Until the early 1990s procurement management had been characterized by a strong domestic bias, with local Asian suppliers playing only a marginal role. Not only were procurement decisions mostly made by the parent company; in many cases, individual product divisions and profit centers were entitled to decide on their own where to procure. All these domestic procurement offices developed strong ties with domestic suppliers; procurement engineers were trained to handle the multilayered networks of Japanese suppliers but had no incentive, and also lacked the relevant expertise, to search for, qualify, and upgrade independent local suppliers in East Asia.

As long as this system prevailed, it was difficult for the local managers of Japanese affiliates in Asia to overcome the decisions made by the home procurement office. Suppose a local supplier was selected and qualified. The part in question then had to be approved by the parent company. This process could take up to nine months. By the time an Asian affiliate got approval, the part was often not needed anymore. That had very negative consequences: it gave rise to structural rigidities in the procurement system and fostered technological conservatism and low speed-to-market.

Japanese firms are now systematically increasing their reliance on international procurement. Take the case of Hitachi, a behemoth whose consolidated sales equal roughly 2 percent of GNP of Japan, the world's second largest economy.[38] Earlier than many of its Japanese rivals, Hitachi internationalized procurement; and it has kept upgrading these activities to changing competitive requirements. In the early 1970s Hitachi was among the first Japanese electronics firms to set up overseas procurement bases in the United States and Hong Kong. In 1989 it established a number of new international procurement offices (IPOs) in San Francisco, Singapore, Seoul, and Taipei. It also began to provide assistance to cut prime costs and train foreign suppliers. However, only after a major reorganization in August 1993 did this process of deepening really get under way. Hitachi established a Center for the Promotion of Procurement in Asia in Singapore. While Hitachi's IPOs for all practical purposes have been commercial purchasing offices staffed primarily by buying agents, in the new center engineers (both from Hitachi and its suppliers) are involved throughout all stages of the procurement decision, including component design and materials specification. The center thus acts as a mechanism for bringing foreign suppliers into Hitachi's internal design processes and for shifting to longer-term supply arrangements.

It is notable that policy incentives provided by both the Japanese government and various host countries in the region have induced Hitachi to rely more on procurement in Asia. For instance, tax incentives for import promotion developed by the Japanese government have helped Hitachi to reduce the cost of importing components from East Asia. The irony is that these incentives, which were originally developed in response to pressures from various U.S. administrations to increase the domestic market share for U.S. companies, have facilitated some overdue organizational adaptations of Japanese firms.

The policies of the host country have also been of great importance. In Malaysia, for instance, Hitachi as much as Matsushita and others have closely cooperated with the government's programs for promoting domestic industries. One example is the Penang Skill Development Center of the Penang Free Export Zone, where Hitachi together with Japanese parts manufacturers participate in training programs for local parts manufacturers.

Redeployment of Japanese Component Production. Important changes have also occurred in the role of Japanese component suppliers: the tight linkages that traditionally bound Japanese electronics firms and their domestic suppliers are losing much of their vigor. Those suppliers that produce relatively complex and higher value-added components have substantially enlarged their investment in East Asia, primarily in Malaysia and Thailand and increasingly also in China. Once these suppliers have established production in East Asia, they are much less inclined to stick to their traditional clients. To amortize as quickly as possible their investment outlays and to gain economies of scale, these affiliates are now actively searching for new clients, with the result that they frequently supply a number of Japanese firms as well as American, Taiwanese, Korean, and some European firms.[39]

Most lower-level Japanese subcontractors are fairly small firms that since the recession started in 1991 have been under tremendous pressure from their customers and parent companies to lower prices. Most of them are reported to have "a terrible time coping with the steps taken by their parent companies to deal with the higher yen,"[40] and have reached the limits to comply with these requests.[41] They are thus "faced with the choice of investing overseas or closing down."[42] Many of these lower-level Japanese component suppliers, however, may not be able to raise the funds required for such investments. The cost of domestically raised investment capital has substantially increased, especially affecting small and medium-size enterprises (SMEs), which, in contrast to the big multinationals, have to rely on bank loans for funding their overseas production. Given such severe financial constraints, most lower-level Japanese component suppliers have not been able to invest on their own in overseas production affiliates. Many of them have gone out of business. The affiliates of Japanese higher level component suppliers thus increasingly have to rely on domestic Asian subcontractors, mostly through various contractual, nonequity arrangements such as consignment production and contract manufacturing.

Shift to Regional Supply Sources. A fourth important consequence of the yen appreciation is that Japanese affiliates in Asia have increased their direct purchases from both Taiwanese and Korean suppliers. Take for instance Taiwan's Tatung group. Tatung's Chunghwa Picture Tube affiliate in Malaysia now supplies a number of Japanese TV-set makers in Malaysia, Singapore, and Thailand.[43] The yen appreciation has also been a driving force behind the spread of OEM and contract manufacturing arrangements with Asian companies, primarily from Taiwan. Since 1994 Japanese PC manufacturers have drastically increased their purchases of PCs, motherboards, terminals and monitors, and a variety of other PC-related products from Taiwanese computer companies. NEC (Nippon Electric Company), for instance, gets monitors and motherboards from Tatung and Elite, and Fujitsu, Epson, Canon, Hitachi, Sharp, and Mitsubishi have all become major OEM customers.[44]

A note of caution must be added. While Japanese affiliates in Asia have increased their reliance on regional and local procurement sources, component imports from Japan have also grown rapidly. One explanation for this apparent contradiction is that the expansion of Japanese production networks in Asia has been so fast that it is compatible both with an increased regional sourcing and with increased component imports from Japan. A second explanation may be that most key components still have to be sourced from Japan or from Japanese firms producing in the region.[45] Still another potential explanation is that as more components are sourced within the region, so more capital equipment must be imported from Japan to produce these components.

Penetrating Asia's Growth Markets: A Shift to a Broader Product Mix

A second important concern is market penetration, especially for companies with a large stake in consumer electronics. Japanese electronics firms are now eager to penetrate the rapidly growing markets of Asia in order to compensate for the slower growth of demand at home and in the United States and Europe. This penetration is no longer restricted to low-end consumer products but includes complex and differentiated products like high-precision components and industrial electronics. Effective market penetration requires a redeployment of production as well as closer links with local firms.

Motivations. Most Japanese electronics companies traditionally focused on the expansion of rapid market share by shortening the product cycle and relying on "product variety wars." Constant product differentiation was the main vehicle for market share expansion. That was a tremendously successful strategy—as long as rapid demand growth could be taken for granted. This is no longer true today. Most electronics markets have shifted from sellers' to buyers' markets,[46] demand for consumer electronics remains muted both in Japan and Europe, and competition for the U.S. electronics market has intensified.

As a result, Japanese electronics firms are now under great pressure to create new product markets. Yet despite continuously high investments into product development, only a few potentially successful high-growth products have emerged.[47] The main emphasis thus has to be on geographic market diversification, especially into the rapidly growing markets of East Asia. That is why Japanese electronics firms today are all anxious to expand their market shares in East Asia. Unlike the second half of the 1980s, when supply considerations were the dominant concern, today Japanese overseas investment in East Asia is driven by "a completely new logic,"[48] and market share expansion within the region is a much more prominent objective. This is especially true for China. Almost all the leading Japanese electronics firms are committed to major new investment projects in the PRC: getting a foot into this potentially huge growth market has been the main motivation for such investments, overriding the substantial concerns about political and macroeconomic instability and the huge investment risks involved.

A Strategic Market for the Electronics Industry. Until quite recently the Triad, consisting of North America, Europe, and Japan, was considered a proxy for the world market. All other countries were lumped together under a residual category called ROW, meaning the rest of the world. This classification is no longer possible. East Asia (exclusive of Japan) today accounts for roughly 22 percent of world GDP, a considerable market size. The share is expected to grow to roughly 35 percent by 2010.[49]

For the electronics industry, East Asia has been a strategic growth market since the late 1980s. High domestic savings ratios and the integration into international production networks has led to extraordinarily rapid growth. Since 1990 most countries in the region have had annual GDP growth rates of between 6 and 8 percent. That has led to the rapid growth of disposable income for urban middle-class people, who are familiar with the latest consumer gadgets and computer wizardry and who are able to pay for them. Pockets of extreme wealth have emerged in all countries of the region, including such "new frontier" countries as China, Indonesia, and Vietnam.

But disposable income has also increased for lower-income groups, with important implications. Japanese electronics firms will be forced to broaden the mix of products they produce within the region. So the complexity of their Asian production networks is bound to increase. That will make it even more difficult for them to control and coordinate these networks in a centralized manner.[50]

Take the example of Sony. The share of Asia (including India, but excluding Japan) in Sony's total sales increased from 6 percent in 1985 to 20 percent in 1995. What really matters, however, is that Asia is, to quote Sony's marketing chief, "a gold mine for *existing* products."[51] Sony expects that Asia's share in its sales of existing products will increase to more than 50 percent over the next few years. In other words, Sony expects Asia to become its main market for homogeneous products, which would enable it to buy time and breathing space and to generate the income required for implementing its diversification strategy. For companies like Sony, as well as for Matsushita and Sanyo, this development could be of critical importance; these companies desperately need a new cash cow that can provide the funds for new product development.[52]

The growth of the Asian market has important implications for the organization of international production. The rapid growth in demand for standard consumer goods implies that moving down-market is a sensible strategy; all Japanese producers of consumer devices in fact have implemented such a strategy. But because of the razor-thin profit margins that characterize most of these products, their production has to be located at low-cost sites and close to their main growth markets, primarily in Asia. Yet new challenges are already emerging. Japanese firms still dominate most markets in the region, but firms based in Korea, Hong Kong, and China have rapidly caught up, and they are now aggressively moving up-market. The result is that Japanese firms have ceased to be able to charge premium prices, and this is true even for new frontier markets like China, Vietnam, and Burma.[53]

A Broader Product Mix: Components and Industrial Electronics. Until the late 1980s most of the Asian production networks of Japanese electronics firms covered only a limited variety of products, markets, and production activities. Over the last few years substantial changes have occurred in this traditional product mix and put new pressures on Japanese electronics firms to open up their Asian production networks. Take, for instance, electronic components. East Asia has now become a major market for many high-precision components such as microprocessors, displays and large-size picture tubes, and components for hard disk drives and computer printers.

In 1993 the value of Japan's exports of electronic components to East Asia was almost $13 billion,* roughly 45 percent of its total electronics exports to the region.[54] For some countries, the share was substantially higher. More than 61 percent of Japanese exports to Taiwan and Korea were components, and nearly 59 percent of exports to Malaysia. In the same year the value of U.S. exports of electronics components to East Asia was almost $10 billion, roughly 55 percent of total U.S. electronics exports to the region. For Malaysia, a major U.S. supply base for IC assembly and PC-related products, this share was 83 percent. Components are also by far the most important product group of U.S. electronics exports to Taiwan; their share has consistently increased since 1985, from more than 31 percent to almost 56 percent.[55] Most of these components are semiconductors that are shipped to Taiwan as inputs to its burgeoning PC industry. The growing share of components in Japanese and U.S. exports to Taiwan clearly reflects the increasing importance of Taiwan as a global supply base for the world PC industry.

The continuous rapid growth of the East Asian market for electronic components is significant. The more these countries proceed to

*All dollar amounts are U.S. dollars unless otherwise indicated.

upgrade their electronics industry, the more they are dependent on component imports. Take one example: both in 1994 and 1995 Japanese exports of components to Asia increased by almost 35 percent.

The shift to a broader product mix now also includes industrial electronics, especially computer-related products and telecommunications equipment. Again one finds the same dual-market structure as for consumer electronics: mass markets for homogeneous products coexist with increasingly important emerging markets for differentiated products. Desktop and laptop PCs are examples of homogeneous products; differentiated products include a variety of equipment required for computer networks (especially PC servers) and specialized work stations.

Computer-related products face a huge and still largely untapped demand potential in this region. Most Asian countries still display very low levels of computerization.[56] Malaysia has only 13.0 computers per 1,000 people, and the share is substantially lower for Indonesia (3.0 per 1,000) and China (2.1 per 1,000). Ranking at the bottom is Vietnam, with a penetration ratio of 0.3 computers per 1,000 people. If one compares these ratios with the region's leader, Singapore,[57] which currently has about 80 computers per 1,000 inhabitants, it becomes clear that the former countries have a huge potential for future market growth. In ASEAN countries (excluding Singapore), annual growth rates of demand for PCs are estimated to be about 40 percent. Most noteworthy is the exploding demand for computers that China has experienced since 1992: in unit terms, sales volume increased from 180,000 units in 1991 to almost 1.7 million in 1995, with annual growth of demand exceeding 50 percent on average.[58]

During the fall of 1992 U.S. firms such as Compaq and Dell, followed later by Apple and others, started a blistering price war in Japan, offering Japanese-language machines for roughly half the price of NEC machines. Most of these machines are actually produced at locations in the China Circle: both Apple and Compaq have production affiliates in Singapore, and both rely heavily on OEM arrangements with Taiwanese computer companies.[59]

That price war posed a serious threat to NEC's dominance of the heavily overpriced Japanese PC market that until then had seemed to be invincible. NEC's immediate reaction was to gather its main subcontractors, just before Christmas 1992, for a secret meeting in rural Gumma Prefecture north of Tokyo, asking them to come up within three weeks with "50 percent cuts across the board, all parts, all assemblies, everything."[60] During the following months it became clear, however, that many of NEC's suppliers had no more leeway to cut costs, and some were even starting to rebel. The traditional response to price competition—that is, shifting the burden of cost reduction onto the shoulders of subcontractors—clearly had reached its limits. As a result, NEC was forced to move, fairly ad hoc and without much preparation, a growing share of its production abroad, especially to lower-cost locations in East Asia.

For its main product line, the PC 9800 series, NEC chose a two-pronged approach. In April 1994 it shifted the design of the motherboard (the main circuit board for PCs that contains the central processing unit) to its Hong Kong subsidiary, NEC Technologies Hong Kong. The main objective was to redesign the board so that it could use more of the cheaper standard components available from Korean, Taiwanese, and Chinese producers.[61] By 1995 NEC had increased the share of these East Asian components to about 70 percent of the board's value. Since about mid-1995 NEC has been assembling an increasing number of these new PC models in its new joint venture in Shanghai, which was originally established to assemble NEC workstations for China's domestic market.

This example clearly shows that, confronted with an increasingly pervasive price war, Japanese computer manufacturers have cast aside most of their earlier inhibitions to forge close ties with Asian suppliers. They are now engaged in a somewhat belated attempt to tap into and replicate the production networks that U.S. computer companies have established in the region.

Most governments in East Asia consider the spread of information technologies to be an essential prerequisite for economic development and are eager to involve foreign computer companies in the development of both IT (information technology) applications and IT production. Import restrictions, however, are pervasive, with the result that market penetration requires local production.[62] As a quid pro quo to improved market access, Japanese firms face increasing pressure to comply with the requirements of host country governments to open up their supplier networks and to localize component sourcing, key management functions, and R&D.

Those changes are especially needed in telecommunications equipment (ranging from fax machines, pagers, cellular phones, to switching and transmission equipment), for which Asia is now the most important growth market. Competition for the markets is extremely intense, with the result that all major manufacturers of telecommunications equipment are expanding their production in the region. To penetrate, for instance, the closed public procurement markets for transmission and switching equipment, foreign firms must establish domestic production. The typical entry strategy into such markets requires the foreign company to accept offset production agreements that allow local firms to participate in production. For most of the necessary support services to install and upgrade telecommunications equipment, it would simply be too costly and time consuming to provide them from abroad. As a result, Japanese firms have had to expand those circuit design and software activities in East Asia that are required for installing telecommunications systems.

These are important new developments. Very little overlapping and rivalry has occurred so far between U.S. and Japanese sourcing strategies: the Americans have focused on PC-related products, while the Japanese have focused on consumer electronics and appliances. That is now rapidly changing, since Japanese firms are shifting many PC-related products to East Asia. Therefore, for the first time, U.S. and Japanese firms will have to compete for the same potential supply sources in East Asia. How will this affect the strategies of U.S. firms? Will they be forced to expand their in-house component manufacturing activities in Asia, as Seagate is doing? Will they be forced to establish centralized control in order to keep a tighter rein on technology leakage that could benefit their Japanese rivals? In other words, will the pendulum now swing back, after a long period of extended outsourcing, to more integrated forms of organizing the international production networks of U.S. computer companies?[63] And how will that affect the approaches of U.S. electronics firms toward their external suppliers and contract manufacturers? Will they be forced to establish longer-term links with local suppliers to gain effective control? Presumably, U.S. firms must act to prevent Japanese firms from developing East Asia into their exclusive supply base in the future.

Mobilizing and Harnessing the Region's Capabilities

Japanese electronics companies are equally attracted by supply-side factors. East Asia's excellence in low-cost manufacturing remains a prime attraction. But the region can now also provide higher value-added support functions such as flexible specialization, engineering, and product and component design. Access to the region's improved capabilities is now considered to be an essential prerequisite for a successful upgrading of Japan's domestic production system. To mobilize and harness these capabilities, Japanese firms are being forced to broaden their capability transfer to East Asia and to internationalize their innovation management.

East Asia's Improved Capabilities. Simple cost considerations are no longer the only factor that attracts electronics firms to this region. Gone are the days when East Asia was just a source of cheap labor for final assembly activities. Since the mid-1980s substantial improvements have occurred in the locational advantages of East Asian production sites. The overall picture is encouraging: investment in infrastructure is booming; every country has increased its efforts to educate and train people across all levels of the occupational ladder; financial systems have been liberalized; domestic firms have deepened their technological capabilities; especially in the export sector, domestic firms have vastly improved their organization and management approaches; and, finally, governments have in general become much more pragmatic in dealing with foreign investment and are attempting to foster closer links between foreign investors and domestic firms.

As a result, East Asia has seen the emergence of specialized centers of expertise in the production and often also in the design of certain products. Some examples are semiconductors in Korea, a variety of PC-related products and components in Taiwan, disk drives, printers, and sound cards in Singapore, and household appliances and consumer electronics in Malaysia and Thailand. Hong Kong and China have consistently concentrated on two product groups: household appliances and consumer electronics. Since 1990, however, the share of computer-related products has rapidly increased. To a large degree, that reflects the rapid expansion of Taiwanese production networks into China, with a focus on motherboards and monitors.[64] Leading electronics firms, both in the United States and Japan, are now eager to capture some of the externalities generated in these geographically concentrated centers of expertise either by forging links with domestically owned firms or by establishing a local subsidiary.

These centers of expertise reflect two distinct, yet complementary, strengths of this region: a capacity to increase at an incredible speed the production of highly capital-intensive and complex mass production lines like monitors, disk drives, and computer memories; and a capacity for quick response to changes in market requirements and technology through flexible specialization in manufacturing and procurement. While Korean *chaebol* like Samsung demonstrate most clearly the first strength, medium-size Taiwanese computer companies like Acer are typical examples of the second strength.[65] The improved capabilities are described by Borrus (chapter 5) and Chung (chapter 6)

A Broader Capability Transfer. Japanese electronics firms are now experimenting with new approaches to innovation management. This may help to improve the ability of Japanese firms to mobilize and harness the region's capabilities. These experiments are driven by the need to outsource different capabilities either that have become too expensive in Japan or that only a few firms can afford to retain. In doing so, Japanese electronics firms seek to emulate successful strategies of American electronics firms: to improve their specialization and hence to strengthen their core competencies through an increasing reliance on outsourcing.[66]

There is a rich literature on the comparative strengths of innovation management by Japanese firms.[67] They have the ability to reduce the development cycle for new products and thus to accelerate speed-to-market, as long as these products remain within a given technology paradigm. A continuous refinement of product design and process engineering have been hallmarks of the Japanese approach to innovation management. However, more recent research, which has been largely conducted by Japanese researchers and is much less well known in the West, has highlighted some important weaknesses of the international innovation management strategies of Japanese electronics firms.[68] Compared with their American and European counterparts, Japanese firms are still at a relatively early stage of R&D internationalization and so far have limited experience in organizing international R&D networks.

Figure 7-1 summarizes some empirical evidence for R&D activities of Japanese electronics firms in East Asia. I distinguish five main categories:

—adaptive engineering, that is, engineering activities that go beyond basic manufacturing support services and include the incremental adaptation and improvement of products and processes;[69]

Figure 7-1. *Japanese Electronics Research and Development Activities in East Asia*[a]

	China				*Hong Kong*				*Korea*				*Malaysia*				*Singapore*					*Taiwan*				*Thailand*			
Firm	*I*	*II*	*III*	*IV*	*I*	*II*	*III*	*IV*	*I*	*II*	*III*	*IV*	*I*	*II*	*III*	*IV*	*I*	*II*	*III*	*IV*	*V*	*I*	*II*	*III*	*IV*	*I*	*II*	*III*	*IV*
Aiwa																				X									
Alpine			X																										
Canon											X											X							
Casio																X													
Fujitsu			X				X											X	X				X						
Hitachi			X											X		X						X							
JVC																	X												
Matsushita													X			X		X	X		X		X	X	X				
Melco													X																
Mitsubishi			X																										
NEC			X				X	X											X				X					X	
Nesic																										X			
Sharp												X				X							X	X					
Sony																X			X			X							
TDK																X													
Toshiba						X				X								X		X			X						
Uniden					X																								
Total			5		1	1	2	1		1	1	1	2	1		6	1	3	4	2	1	3	5	2	1	1		1	
Country total	5				5				3				9				11					11				2			

Source: Berkeley Roundable of the International Economy-FDI database, based on news reports.

a. I. Adaptive engineering (total 8); II. IC design (total 11); III. software engineering (total 15); IV. product development (total 11); V. generic technology development (total 1).

— integrated circuit design;

—software engineering (which ranges from simple program reconversion to fairly sophisticated projects);

— product development, most of it for the local market, and a few projects for the regional market; and

— generic technology development, that is, major innovations with a huge potential for productivity enhancement and the creation of new product markets.

The figure shows that out of a total of forty-five projects, only one falls under the last category, that is, the audiovisual information research center of Matsushita in Singapore, established in 1990, which focuses on the development of compression technology for image transmission.

The largest share of Japanese electronics R&D activities in East Asia falls under two categories: software engineering (with fifteen cases) and circuit design (with eleven cases). The essential point to stress is that in most instances both are essentially support services required to enter or expand the region's domestic markets. For software engineering, for example, the development of Chinese language programs plays an important role, with the goal to improve the market position in China for Japanese computer manufacturers. And most of the circuit design activities are dedicated to application-specific integrated circuits (ASICs) that are required for consumer devices or telecommunications equipment sold in the domestic or regional markets. Both Singapore and Hong Kong have recently emerged as regional IC design centers for consumer devices. Japanese firms are now concentrating their limited resources at home on higher value-added products related to computing, multimedia, and networking applications, and thus are eager to redeploy design and engineering functions linked to audio-visual equipment and home appliances. Take the example of Sharp's IC design center in Singapore, which currently consists of ten people.[70] Apart from after-sales support services, its main function now is to program microcontrollers embedded in home appliances. Over the next years Sharp intends to use this center to outsource design work from Japan on a much larger scale. The company is planning to increase the center's staff to thirty people by 1998.

The main driving force for relocating R&D activities to East Asia is the current shift from proprietary components to standard components that can be sourced at lower cost from local or regional suppliers. To achieve this goal, Japanese electronics firms have all been forced to upgrade their regional and local support services. A second important motivation for Japanese electronics firms to expand their R&D activities in Asia is to tap into existing pools of lower-cost human resources. Most countries of the region, in fact, pursue aggressive policies to increase the supply of engineers and scientists. As a result of the closed production networks that Japanese electronics firms established during the 1980s, they are facing much greater difficulties than U.S. companies in attracting top local engineers, and they are under great pressure to improve recruiting.[71] To do so will not be easy. Japanese affiliates in Asia find it difficult to retain trained foreign personnel not only because of the glass ceiling that normally prevents foreigners from reaching top management positions but also because of the relatively slow pace of upward mobility for managers and engineers.[72]

After some unsuccessful attempts to hire engineers by paying higher salaries, the Japanese affiliates have now developed a peculiar recruitment approach that builds on some inherent strengths of the Japanese production system.[73] They now hire most of these local engineers internally. Using a careful selection process, each affiliate develops a pool of highly motivated technicians, whom they then train for five to seven years to become (possibly unlicensed) engineers. That accounts for the relatively low turnover at the engineering level—the new engineering skills are firm specific. This peculiar recruitment approach obviously builds on existing strengths of the domestic Japanese production system. At the same time it allows Japanese firms to overcome two problems: most host countries limit the immigration of Japanese engineers; and there is an implicit understanding among Japanese firms that bidding up salaries to attract engineers should be avoided.

Japanese firms now also attach much greater importance to market intelligence and product customization. They are increasingly aware that Asia is characterized by heterogeneous demand patterns and highly segmented product markets. At the same time the variety of production sites has kept increasing, so that Japanese firms must adapt their Asian production networks to the idiosyncrasies of each of these markets. As a result, local affiliates need to have a capacity for continuous product customization, that is, they need to establish on the spot a capacity for continuous redesign (adaptive engineering). Adaptive engineering and some development activities have become increasingly decentralized and take place in engineering departments of Asian manufacturing affiliates.

A Shift toward Decentralized Governance Structures

Japanese production networks shifted to highly centralized governance structures with limited local roots during the creation of export platforms in the late 1980s. However, the pendulum is now swinging back toward decentralization and local embeddedness. Asian production networks are now much more complex, owing to the increased share of high value-added support services and the broader mix of products produced. The task of reconciling the conflicting requirements of export platform production and regional market penetration is difficult to accomplish at a distance. This has led to a gradual shift to more decentralized governance structures, in which regional headquarters and individual Asian affiliates take over many coordination functions that used to be the privilege of the parent company.

Consequently, OEM purchases, subcontracting, and contract manufacturing arrangements have expanded rapidly. The most important benefit is to speed up decisionmaking. For most of the products produced in Asia, time-to-market has become the most important determinant of success. Japanese affiliates in Asia thus cannot wait the roughly nine months that are normally required to get approval on procurement from headquarters. Each of the individual nodes of Japanese Asian production networks now have to contribute to the lead company's resources and capabilities.

Until the early 1990s most of the funds required for the expansion of Japanese regional production activities in East Asia came from remittances from the parent company in Japan. Reinvestments of overseas affiliates and equity links with local investors played a minor role.[74] Today most of these investments are locally funded. Between 1989 and 1992 the ratio of reinvestments of Japanese affiliates to Japan's total foreign direct investment increased from 35 percent to 60 percent for ASEAN affiliates, and from 54 percent to 80 percent for NIE (newly industializing economy) affiliates. These figures are well above the ratios reported for the United States, which increased from 15 percent (1989) to 24 percent (1992), and were made possible by the high profitability of Japanese affiliates in Asia. In fiscal 1992 the ratio of ordinary profit to sales of Japanese overseas manufacturing affiliates was 5.1 percent in ASEAN and 5.6 percent in Asian NIEs, in sharp contrast with minus 0.2 percent for Japanese affiliates in the United States.

These data have important implications for the governance structure of Japanese Asian production networks. To the degree that Japanese affiliates are now much less dependent on their parent companies for investment funds, they are also now gaining more scope for local decisionmaking. Moreover, Japanese affiliates in Asia, have now begun, at long last, to tap into the region's thriving equity markets and are listed on local stock exchanges.[75] Sooner or later these changes will gradually broaden the scope for local decisionmaking.

Implications for the China Circle

Japanese and China Circle networks are likely to interact more closely in the future, and that may greatly change the dynamics of competition in the electronics industry.

The Role of China

A regional specialization is now beginning to emerge for Japanese networks in East Asia. In the electronics industry, one can distinguish roughly the following pattern: Singapore and Hong Kong compete for a position as regional headquarters (together with major support functions like procurement, testing, training, engineering services, and some product design); South Korea and Taiwan compete for OEM contracts (including some design activities) and as suppliers of precision components; Malaysia and Thailand, and now also the Philippines, are preferred locations for the volume production of, in particular, midlevel and some higher-end products.

China's role is critical. Much depends on how Japanese networks are going to integrate this huge, quasi-continental economy that is now also becoming a major geopolitical force. China's overwhelming attraction is its potentially huge domestic market for a wide range of electronics products. It also now competes as a new export platform production base for low-end assembly and simple components manufacturing.

Market Access

First it is important to look at the issue of market access and how it has shaped China's integration into Japanese production networks. Japanese firms so far have concentrated almost exclusively on two market segments: consumer electronics (including household appliances) and electronic components. Japanese firms have failed to play any significant role in China's rapidly growing markets for industrial electronics, except in telecommunications equipment.[76] That is true particularly for computer-related products, which are dominated by U.S. and Taiwanese firms.[77] It is important to note that most of the computers sold with a U.S. label were produced as OEM products in the China Circle, primarily in Taiwan but also in Singapore. Fundamental differences thus exist in the logic of integration of China with U.S. and Japanese production networks. For U.S. networks, the focus clearly is on computer-related products and on telecommunications equipment, while for Japanese networks the focus overwhelmingly has been on consumer products.

China's integration into Japanese production networks for consumer electronics merits a closer look. China's market potential for electronics products is mind-boggling. Seventy percent of China's population of 1.2 billion people, that is, 840 million people, live in the countryside. Of these 840 million people, only 10 percent own a color TV set, compared with almost 90 percent of the Chinese urban population. Market penetration is substantially lower for more sophisticated consumer goods. VCRs, for example, have spread to only 3 percent of the Chinese population. As a result of this huge untapped market potential, all the leading Japanese and Korean TV makers have aggressively invested in domestic market–oriented production facilities. Competition has intensified to such a point that pervasive price wars have drastically reduced the profit margins for most of these affiliates. The massive wave of recent investment projects that went to China in consumer electronics has also given rise to a serious long-term problem: it may actually slow down progress to an improved regional specialization of Japanese Asian production networks. In some cases these investments in China have led to substantial surplus capacities within the region, both for final assembly lines and component manufacturing.

Take the case of the Matsushita group, which since 1992 has aggressively expanded its China presence and now has nineteen affiliates in that country.[78] So far, Matsushita's Chinese affiliates play a secondary role for its Asian production network: the production value of its China affiliates accounts for less than 5 percent of Matsushita's total international production. The largest chunk is still generated in the ASEAN countries, where an estimated 60 percent of Matsushita's total international production value of $6 billion originates. The goal now is to raise China's share to roughly 20 to 25 percent of Matsushita's international production value by the year 2000. This goal has given rise to notable structural adjustment problems. Matsushita's joint venture in Beijing for TV picture tubes illustrates those problems.[79] That project is widely considered to be a major test case for Japanese FDI in China, and it has received privileged treatment. Nevertheless, serious problems have emerged. Since local supplier industries are still weak, 15 percent of the components used (all of them key components) have to be imported from Japan. Because of the appreciation of the yen, the price for these components has increased rapidly, leading to a severe profit squeeze. In addition, labor costs have risen dramatically and now approach Thai wage levels, while productivity continues to lag behind Thai productivity levels. Probably the most serious problem, however, has resulted from unexpected limits to the growth of the domestic market. Demand for lower-end TV sets is already reaching saturation: almost 90 percent of China's urban households are now estimated to have such TV sets. Furthermore, roughly one-fifth of the 13 million TV sets that are sold per year in China are smuggled across the borders, many of them produced, ironically, by Japanese affiliates in Southeast Asia.

The result is that Matsushita must radically change its strategic focus. Instead of aiming primarily at the domestic market, which was the original motivation, it must now increase the share of production that goes into exports. Doing so will probably generate serious problems of overcapacity. This comes on top of structural adjustment problems that still cry out for appropriate solutions, the result of the earlier wave of investments in ASEAN during the 1980s. One key feature of the Asian production network of Matsushita is the coexistence of mini-Matsus oriented toward the domestic market with more recent export bases in the same countries. This is causing serious conflicts of interest: similar products are often produced at both facilities, but at very different productivity and quality levels. Different facilities have also generated different types of sourcing arrangements. In principle, plants of very different vintages can coexist competitively as long as they are producing different qualities for different market segments at different prices. Low-productivity production for the domestic market can be profitable if the domestic market remains highly protected. That, however, is no longer true, since domestic markets for consumer electronics are now gradually being opened up to international competition.[80]

Following on the first wave of Japanese investment in East Asia, between 1986 and 1990, the new wave of investment in China has generated substantial surplus capacities in the region. Sustaining such a dual production structure is a costly proposition. Yet overcoming it may not be easy, because of the heavy "sunk investments." Companies like Matsushita thus can no longer postpone decisions on where to consolidate their individual regional supply base for particular components, such as TV picture tubes.

China's Role as an Export Platform Production Base

This brings me to the second aspect of China's integration into Japanese Asian production networks. China so far competes primarily with Indonesia, India, and now also Vietnam as a new export platform production base for low-end assembly and simple component manufacturing.[81] But that is beginning to change. China may emerge as an alternative site to Malaysia and Thailand as Japanese firms aggressively search for ways to overcome the dual-production structure and the accumulated surplus capacities in Asia. They may decide to close down production lines in those two ASEAN countries and to redeploy them progressively to the much larger Chinese market. In response to this threat of a possible investment diversion to China, the Malaysian government as well as other ASEAN governments are now willing to accelerate trade liberalization through the Asian Free Trade Area, primarily "to create a market of 350 million people that can draw foreign investment and compete on a more level playing field with China."[82]

All this implies that formidable barriers still need to be removed before the potential for regional specialization can be fully realized for Japanese Asian production networks. The massive wave of recent investments of Japanese consumer electronics firms into China has undoubtedly created serious structural adjustment problems that may not be easy to solve in the short run. This activity has further increased the pressure on Japanese firms to rationalize their Asian production networks, that is, to improve their regional specialization. It is important to note, however, that Japanese firms certainly cannot dictate unilaterally the necessary changes; they now need to find a delicate balance between the requirements of the China market and the requirements of the ASEAN market. Again that shows why pressures to open up and decentralize Asian production networks are likely to remain a permanent challenge for Japanese electronics firms.

New Opportunities and Challenges for Taiwan

Over the last decade Taiwan has established itself as a world-class supply source for different electronic hardware products; it is the world's largest supplier of computer monitors, motherboards, switching power supplies, mouse devices, keyboards, scanners, and various add-on cards. Since 1994 Taiwan also has become the world's largest manufacturer of notebook PCs. Most of these computers are sold to U.S. and Japanese computer companies, which then resell these machines under their own logo. What matters, however, is that 70 percent of the computers that are sold under such OEM arrangements have been designed by Taiwanese companies.[83] This clearly indicates that Taiwanese computer firms have been able to develop significant design capabilities.

Progress has also been impressive in the field of components. Taiwan today has hundreds of passive component makers that have established a strong position relative to their erstwhile leading Japanese and U.S. competitors. And though Taiwan's semiconductor industry now accounts for hardly more than 3 percent of the world market, some of its firms have developed a strong position for a number of higher value-added IC devices, like chip sets, static RAM memories, mask ROMs, and EPROMs. In addition, Taiwan has one of the world's leading silicon foundry companies, Taiwan Semiconductor Manufacturing Corporation (TSMC), which is able to produce leading-edge ICs for major international semiconductor firms, with very short production cycles and with the most sophisticated process technologies and production equipment.

A second important feature of Taiwan's computer industry is that SMEs have been the main carriers of its rapid development. These SMEs have acted as highly successful carriers of international market share expansion, first through exports and increasingly also through international production. The rapid expansion of overseas production by Taiwanese computer companies, well before they were able to consolidate and upgrade their domestic production activities, has been pioneered by SMEs (see chapter 6).[84]

This expansion runs counter to some well-established beliefs about what a company needs to survive in the international arena, beliefs based on international trade and investment theories. That being small disqualifies a company as a carrier of internationalization is widely regarded as self-evident. Small firms have limited resources and capabilities and thus are unlikely to possess substantial proprietary assets. They also have a limited capacity to influence and shape markets, market structure, and technological change. Small size can therefore act as a powerful barrier to internationalization.

Elsewhere I have shown that the Taiwanese were able to bypass these size-related barriers to international production primarily because they participated early on in a variety of subcontracting, contract manufacturing, and OEM relationships with leading foreign computer companies.[85] These linkages first emerged in consumer electronics with international production networks established by Philips and a number of Japanese companies (especially Matsushita, Toshiba, and Sanyo). Since 1987 Taiwanese firms have become major OEM suppliers for U.S. computer companies. Over the last few years linkages with Japanese production networks have again intensified, this time primarily for computer-related products and components. This clearly reflects some of the fundamental changes in these networks that are analyzed in this chapter.

Effect of the Growing Interaction with Japanese Production Networks

Despite all its achievements, Taiwan's electronics industry is still based on a weak foundation. For most of the key components that determine the price and the performance features of its major export products, Taiwan continues to rely heavily on imports, primarily from Japan.[86]

Take cathode ray tubes (CRTs) for computer monitors. Taiwan's success in the monitor industry has come at a heavy cost: nearly two-thirds of the CRTs that go into these monitors have to be imported, either from Japan or from Japanese affiliates in Southeast Asia. As a result, in 1994 Taiwan had to pay $1.35 billion for CRT imports, making CRTs the largest item of imports from Japan.

The situation is equally severe for display panels, a key component for Taiwan's thriving portable PC industry. For this industry, the ability to purchase display panels in the necessary quantities, at the right time and at a reasonable price, will decide its competitive success. Taiwan has to import virtually all the high-end flat panel displays that are used in its portable PCs, and the supply of these devices is controlled by a tightly knit oligopoly consisting of Sharp, a Toshiba-IBM joint venture, and NEC, with Hitachi and Matsushita being important second-tier producers. In 1993 Taiwan had to spend $500 million on imports of liquid crystal displays (LCDs), with more than $350 million alone spent on advanced TFT-LCDs. Over the last two years prices for these devices have rapidly increased, so that these imports are likely to have become even more expensive.

This heavy dependence on component imports from Japan has been the root cause of Taiwan's exploding electronics trade deficit with Japan. In 1993 Taiwan's trade deficit in components ($2.5 billion) was responsible for almost 72 percent of Taiwan's total electronics trade deficit with Japan.[87] Obviously, that deficit is a critical barrier to a further upgrading of Taiwan's electronics industry.

Yet Taiwan's growing interaction with Japan's Asian production networks is creating a more complex picture, with hard to predict consequences. Two important developments have recently shaped these relationships. First, as shown earlier, Japanese computer companies have drastically increased their OEM purchases from Taiwanese firms since 1994. Second, Taiwan is now also emerging as a critical supply base for a variety of electronics components. Take, for instance, semiconductors. In April 1994 a large Japanese purchasing mission to Taiwan, the first of its kind, ordered $60 million of semiconductor products. One year later, in April 1995, a second much larger Japanese purchasing mission ordered more than double that amount—almost $130 million.[88]

As a result, Taiwan has experienced a dramatic growth of computer-related exports to Japan. From less than NT$3.9 billion in 1990, these imports increased to more than NT$21 billion in 1994.[89] Two products predominated in 1993: components with 34 percent and electronic data processing (EDP) with almost 33 percent. Together they account for more than two-thirds of Japanese imports from Taiwan. Especially impressive was the reported rise of imports between 1994 and 1996: Taiwanese exports to Japan increased by 347 percent for PCs, 169 percent for components, 122 percent for terminals, and 110 percent for monitors.[90] During the same period Taiwan's imports from Japan *declined* by 36 percent for components and 35 percent for other peripherals.

Both countries now have powerful vested interests in expanding bilateral trade links. The April 1995 purchasing mission brought together twenty of the most powerful Japanese electronics firms and the sixteen leading Taiwanese producers of printed circuit boards, semiconductors, and other components.[91] Trade links are now secured by effective institutional arrangements in both countries. In Japan, the main driving force has been the User's Committee of Foreign Semiconductors (UCOM) of the powerful Electronics Industry Association of Japan. Ironically, UCOM was originally established under pressure from the U.S. government, which was hoping that the organization could help to dramatically increase the Japanese market share of American IC producers. In Taiwan, the main driving force has been the Sino-Japanese Economic and Trade Foundation, a private organization founded to reduce Taiwan's bilateral trade deficit with Japan, but one that has strong backing from the Taiwanese government.

Conclusion

It remains to be seen whether the improvement in Taiwan's trade links with Japan will be sustainable. No doubt Japanese PC vendors perceive OEM purchases as an intermediate solution: it enables them to quickly discontinue lower value-added production activities at home. It also enables them to gain time until they have been able to set up their own supply base for some of these products in China and Southeast Asia. This suggests that such rapid growth of OEM contracts is unlikely to last. Some Japanese PC makers have now started to move some production in-house, taking it away from Intel and Taiwan contract manufacturers.[92]

For instance, Fujitsu, Hitachi, and Toshiba are ramping up PC output in their own Japanese plants and cutting back or eliminating their relabeled OEM purchases from offshore contract companies. Both Fujitsu and Hitachi were using Taiwan's Acer to make PCs that they sold in Japan under their own labels. That was obviously a response to the gradual depreciation of the yen, which started in 1995. The move back to Japan is driven by a perception that it is now again possible to use integrated circuits and components from their domestic sister divisions.

This example shows that nothing is automatic about the benefits from participating in international production networks. Periodically there may be important reversals in the distribution of such benefits. The question is how long Japanese PC producers will be able to sustain this relocation of production back home. There are obviously short-term benefits for their component divisions. But whether relocation back home provides a long-term solution remains doubtful. For instance, it is unlikely to facilitate the expansion of international market share. After all, a great advantage of Taiwanese OEM suppliers has been the incredible speed with which they are able to respond to changes in markets and technology. Japanese vertically integrated electronics giants are not noted for such flexibility.

There is thus considerable uncertainty about future trends. However, there is less doubt about the fundamental economic trends that are driving development. Japanese electronics firms have been buffeted in recent years by powerful economic forces. On balance, those forces are strongly pushing Japanese firms toward more open production networks and especially toward more collaboration with China Circle production capabilities. Thus while rivalry among Japanese and U.S. and China Circle producers will continue, cooperation between and interpenetration of Japanese and China Circle production networks will increase. That will create important new opportunities for the China Circle and may create new challenges to the electronics industries in the United States and Europe.

Notes

1. Ministry of Finance (MOF) figures, as quoted in *JETRO White Paper on Foreign Direct Investment* (Tokyo: Japan External Trade Organization, March 1995), p. 17.

2. Over time, the focus of such investments has shifted twice: first from Northeast Asia (Korea, Taiwan, and Hong Kong) to Southeast Asia (primarily Singapore, Malaysia and Thailand); and then, since about 1992, from Southeast Asia to China primarily. While in sheer numbers the shift to China clearly dominates, Japanese electronics firms have also expanded their production networks into Indonesia (since about 1990) and India, as well as now into Vietnam, Burma, and the Asian republics of the former Soviet Union.

3. The concept of an "international production network" is an attempt to capture the spread of broader and more systemic forms of international production that cut across different stages of the value chain and that may or may not involve equity ownership. This concept allows one to analyze the globalization strategies of a particular firm with regard to the following four questions: (1) Where does a firm locate which stages of the value chain? (2) To what degree does a firm rely on outsourcing, and what is the importance of interfirm production networks relative to the firm's internal production network? (3) To what degree is the control over these transactions exercised in a centralized or in a decentralized way? And (4) how do these different elements of the international production network hang together? For details, see Dieter Ernst, "Networks, Market Structure, and Technology Diffusion: A Conceptual Framework and

Some Empirical Evidence," report prepared for the OECD secretariat (Paris: Organization for Economic Cooperation and Development, 1992); Dieter Ernst, "Network Transactions, Market Structure, and Technology Diffusion: Implications for South–South Cooperation," in Lynn Mytelka, ed., *South-South Cooperation in a Global Perspective* (Paris: OECD, 1994); and Dieter Ernst, "Carriers of Regionalization? The East Asian Production Networks of Japanese Electronics Firms," Working Paper 73 (University of California, Berkeley: Berkeley Roundtable on the International Economy, November 1994). (Hereafter BRIE.) A revised and updated version of that working paper will appear as chapter 2 in Michael Borrus, Dieter Ernst, and Stephan Haggard, eds., *Rivalry or Riches: International Production Networks in Asia* (Cornell University Press, forthcoming); and Dieter Ernst, "From Partial to Systemic Globalization: International Production Networks in the Electronics Industry," report prepared for the Sloan Foundation project on Globalization in the Data Storage Industry, University of California, San Diego, Graduate School of International Relations and Pacific Studies, 1996.

4. Three of these newly emerging production networks are of particular importance: those established by Korean *chaebol;* those established by Taiwanese producers of PC-related products and components in Southeast Asia and China; and those networks of suppliers of parts and subassemblies based in Hong Kong, Taiwan, Singapore and mostly owned by ethnic Chinese. For an analysis of some of these newly emerging Asian production networks, see Dieter Ernst, "How Small Enterprises Internationalize: The International Production Networks of Taiwanese Electronics Companies," a study prepared for the U.S.-Japan Friendship Commission Project on Asian Production Networks, 1996; and Kim Youngsoo, "Technological Capabilities and Samsung Electronics' International Production Network in Asia," in Ernst, Borrus, and Haggard, *Rivalry or Riches*.

5. For a presentation of this argument, see the chapter by Michael Borrus in this book.

6. For an excellent theoretical treatment, see Nonaka Ikujiro and Takeuchi Hirotaka, *The Knowledge Creating Company* (Oxford University Press, 1995).

7. Governance describes how, within any particular firm or international production network, control and coordination is exercised and by whom. It consists of common methods and procedures that shape the behavior of network nodes, such as budgetary rules and procedures; evaluation procedures; personnel management practices; and database management, quality control norms, and so forth. Network nodes can be equity-owned affiliates and those legally independent firms that participate in the core company's interfirm networks.

Governance structures can be centralized or decentralized, and they can be strong or weak. A centralized governance structure means that the core company (the network center) exercises full control over all network activities. A decentralized governance structure means that individual network nodes have some decision autonomy. Strong governance means that control and coordination are enforced over a broad range of value-chain activities and are shared by a large number of network nodes. Weak governance, in turn, means that control and coordination can be enforced only over a limited range of activities and that only a limited number of network nodes are subordinated to them. I have used the following indicators to establish whether a governance structure is centralized or decentralized: ownership patterns, the degree of local decision autonomy, local linkages, the role of regional headquarters, and the degree of local funding through reinvestments and a listing on local stock exchanges. For details, see Dieter Ernst, "Globalization, Convergence, and Diversity: The Asian Production Networks of Japanese Electronics Firms," forthcoming in Borrus, Ernst, and Haggard, eds., *Rivalry or Riches*.

8. See, for example, W. Dobson, "East Asian Integration: Synergies between Firm Strategies and Government Policy," University of Toronto, Centre for International Business, 1995, p. 38; Edward Graham and Naoko T. Anzai, "Is Japanese Direct Investment Creating an Asian Economic Bloc?": Institute for International Economics, Washington, 1994; and Peter A. Petri, "The Interdependence of Trade and Investment in the Pacific," in Edward K. Y. Chen and Peter Drysdale, eds., *Corporate Links and Foreign Direct Investment in Asia and the Pacific* (Sydney: Harper Educational, 1995).

9. Philippe Lasserre and Hellmut Schuette, *Strategies for Asia Pacific* (New York University Press, 1995), p. 176. See also James C. Abegglen, *Sea Change: Pacific Asia as the New World Industrial Center* (Free Press, 1994).

10. Matsushita Electric Industrial Company is the world's largest manufacturer of consumer electronics and household appliances. Since the mid-1980s, however, it has diversified aggressively into communications and factory automation equipment, semiconductors, and video software, with the result that today the share of industrial electronics and electronic components in its overall sales is roughly equal to the share of consumer and household goods.

11. Typical examples are Kenichi Imai, "Patterns of Innovation and Entrepreneurship in Japan," paper presented to the Second Congress of the International Schumpeter Society, Siena, Italy, 1988; and Kenichi Imai and Yasunori Baba, "Systemic Innovation and Cross-Border Networks: Transcending Markets and Hierarchies to Create a New Techno-Economics System," in OECD, *Technology and Productivity: The Challenge for Economic Policy* (Paris, 1991).

12. The first of these mini-Matsus was a minority joint venture with a local partner in Thailand, established in December 1961. That was followed, during 1962 and 1963, by two joint ventures in Taiwan and one in Indonesia.

13. MELCOM is a publicly listed company at the Kuala Lumpur stock exchange, with Matsushita holding a 43 percent equity share.

14. They were either SKD (assembly of prefabricated kits, with no circuit board done by affiliate) or CKD (parent company provides all materials and components, but circuit board assembly is done by affiliate). Author's interview at MELCOM, June 1984.

15. Author's interview at Matsushita headquarters, November 1993.

16. Author's interview at Sanyo headquarters, November 1993.

17. For an excellent analysis of the rationale that has induced Japanese electronics firms to postpone overseas production, see Dennis Tachiki and Akira Aoki, "Foreign Direct Investment: Japanese Global Strategies in the Asia-Pacific and the EC," *RIM: Pacific Business and Industries* (Sakura Institute of Research, Tokyo), vol. 4, no. 14 (1991), pp. 23–33; and Shigeki Tejima, "Toward More Open Corporate Strategies: Will Japanese Firms Take Those Strategies to East Asia?" Export-Import Bank of Japan, Research Institute for International Development, Tokyo, May 1996.

18. From about 240 yen to the U.S. dollar in the fall of 1985, the yen's exchange rate experienced a breathtaking decline to below 80 yen to the dollar in the summer of 1995 before rising again to about 110 in the first half of 1996.

19. Edith T. Penrose, *The Theory of the Growth of the Firm* (Oxford University Press, 1959); Richard Nelson and Sidney G. Winter, *An Evolutionary Theory of Economic Change* (Harvard University Press, 1982); Franco Malerba and Luigi Orsenigo, "The Dynamics and Evolution of Industries," *Industrial and Corporate Change,* vol. 5, no. 1 (1996), pp. 51–87; Bengt-Åke Lundval, "Innovation as an Interactive Process: From User-Producer Interaction to the National System of Innovation," in Giovanni Dosi and others, eds., *Technical Change and Economic Theory* (London: Pinter, 1988); Bengt-Åke Lundvall, ed., *National Systems of Innovation: Towards a Theory of Innovation and Interactive Learning* (London: Pinter, 1992); Bo Carlsson and R. Stankiewicz, "On the Nature, Function and Composition of Technological Systems," *Journal of Evolutionary Economics,* vol.1, no. 2 (1991), pp. 93–118; Jens F. Christensen, "Innovative Assets and Inter-Asset Linkages: A Resource-Based Approach to Innovation," *Economics of Innovation and New Technology,* vol. 4 (1996), pp. 193–209; Dominique Foray and Bengt-Åke Lundval, "The Knowledge-Based Economy: From the Economics of Knowledge to the Learning Economy," in OECD, *Employment and Growth in the Knowledge-Based Economy* (Paris, 1996); Nicolai Foss, "Capabilities and the Theory of the Firm," DRUID Working Paper 96-8 (Copenhagen Business School, June 1996); Patrick Llerena and Ehud Zuscovitch, "Innovation, Diversity and Organization from an Evolutionary Perspective: Introduction and Overview," *Economics of Innovation and New Technology,* vol. 4, no. 2 (1996), pp. 180–92; and Peter Maskell, "The Process and Consequences of Ubiquification," papers prepared for the DRUID workshop, Copenhagen Business School, Department of Industrial Economics and Strategy, January 1997.

20. John Stopford, "Building Regional Networks: Japanese Investments in Asia," London Business School, 1995, quotations from pp. 2 and 16.

21. For instance, recent research on learning processes in international production networks has shown that the more dispersed such networks are, the more difficult it becomes to impose standard procedures and exercise centralized control. For an interesting theoretical approach, see David G. McKendrick, Suzanne K. Stout, and Michael T. Pich, "Network Learning in the Development and Transfer of Technology in Multinational Corporations," University of California, Berkeley, Haas School of Business, November 1994.

22. For a detailed discussion, see David Levy, "CCT's International Supply Chain," Harvard Business School Case Study, 1994; and David Levy, "International Sourcing and Supply Chain Stability," *Journal of International Business Studies,* vol. 2 (Second Quarter, 1995), pp. 343–60, and the literature cited in both articles.

23. For a case study on Intel's affiliate in Penang, see Ernst, "From Partial to Systemic Globalization."

24. *White Paper on International Trade and Industry* (Tokyo: MITI, 1994), p. 6.

25. Dieter Ernst and Paolo Guerrieri, "International Production Networks and Changing Trade Patterns in East Asia: The Case of the Electronics Industry," BRIE, and University of Rome, La Sapienza, 1996, to appear in *Asian Economic Journal* (forthcoming).

26. For detailed evidence, see Ernst, "Globalization, Convergence, and Diversity"; and Ernst, "Carriers of Regionalization?"

27. Similar findings are reported in Dennis Tachiki, "Extending the Human Capital Formation Process: Transnational Corporations and Post-Employment Training," *RIM:*

Pacific Business and Industries, vol. 1, no. 23 (1994); and M. W. Sedgewick, "Does Japanese Management Travel in Asia? Managerial Technology Transfer and Japanese Multinationals," paper presented at MIT-Japan Program conference on Does Ownership Matter? Japanese Multinationals in Asia, Cambridge, Mass., September 1995; and United Nations Conference on Trade and Development, *World Investment Report,* 1995 (Geneva, 1995).

28. For evidence, see T. Abo, ed., *Hybrid Factory: The Japanese Production System in the United States* (Oxford University Press, 1993); Dennis Encarnation, *Rivals beyond Trade: America versus Japan in Global Competition* (Cornell University Press, 1992); Dennis Encarnation, "Asia and the Global Operations of Multinational Corporations," Harvard University, Kennedy School Pacific Basin Research Program, 1996; JETRO, *The 9th Survey of European Operations of Japanese Companies in the Manufacturing Sector* (Tokyo, 1993); Martin Kenney and Richard Florida, *Beyond Mass Production: The Japanese System and Its Transfer to the U.S.* (Oxford University Press, 1993); and M. Gittelman and Edward Graham, "The Performance and Structure of Japanese Affiliates in the European Community," in Dennis Encarnation and Mark Mason, *Does Ownership Matter?* (Oxford: Clarendon Press, 1994).

29. The emerging regional specialization of Japanese Asian production networks with special reference to the China Circle is discussed later.

30. Data provided by the Electronics Industries Association of Japan, June 1995.

31. Though an impressive increase, a 10 percent import ratio still indicates a low domestic market penetration. The exception is color televisions, with a 35 percent import penetration in 1993. Since 1993 Japan has imported more TV sets and audio equipment than it exports. Figures on import penetration from JETRO, *Nihon no seihin yunyu 1993* (Japan's manufactured goods imports in 1993) (Tokyo, 1994). Import penetration is defined as imports divided by (production minus exports plus imports) times one hundred.

32. Asia's share was highest for imports of consumer electronics and appliances: 84 percent in 1993. Kenichi Takayasu and Yukiko Ishizaki, "The Changing International Division of Labor of Japanese Electronics Industry in Asia and Its Impact on the Japanese Economy," *RIM: Pacific Business and Industries,* vol. 1, no. 27 (1995), pp. 2–21. According to Shujiro Urata, "Emerging Patterns of Production and Foreign Trade in Electronics Products in East Asia: An Examination of the Role Played by Foreign Direct Investment," Asia Foundation, conference on Competing Production Networks in Asia, San Francisco, April 1995, table 7, the share of sales to Japan as a proportion of total sales of affiliates of Japanese electronics firms in ASEAN countries increased from 17.5 percent in 1989 to almost 28 percent in 1992.

33. Information provided by the Electronics Industry Association of Japan, January 1996.

34. Based on telephone interview with Fujitsu, May 22, 1996.

35. Fujitsu has relied primarily on OEM supplies from Acer of Taiwan.

36. The survey covered Japanese manufacturers capitalized at more than $100 million that have overseas production affiliates, with 219 companies responding. For details, see JETRO, *Nihon kigyo no kaigai jigyo tenkai no jittai chosa* (Survey of overseas business of Japanese corporations) (Tokyo, 1994).

37. Shigeki Tejima, "Toward More Open Corporate Strategies: Will Japanese Firms

Take Those Strategies to East Asia?" Export-Import Bank of Japan, Research Institute for International Development and Development, Tokyo, May 1996, pp. 3, 16.

38. The following is based on interviews with Hitachi in November 1993.

39. Acer Peripherals' monitor plant in Penang-Malaysia, for instance, sources most of its cathode ray picture tubes from Hitachi and Matsushita affiliates in the region. Interview with K. Y. Lee, president of Acer Peripherals, December 1994.

40. Central Bank for Commercial and Industrial Cooperatives, *Survey of the Effects of the Yen Appreciation on Small and Medium-Size Enterprises* (in Japanese) (Tokyo, 1993), p. 5.

41. For evidence, see JETRO, *Survey of Overseas Business of Japanese Corporations,* p. 18; and Kenichi Miyashita and David W. Russell, *Keiretsu: Inside the Hidden Japanese Conglomerates* (McGraw-Hill, 1994), p. 199 and passim.

42. JETRO, *Survey of Overseas Business of Japanese Corporations,* p. 20.

43. Interviews at Tatung group secretariat, April 1993 and December 1994.

44. This new wave of OEM contracts has had important implications for Taiwan's electronics industry that I will discuss later in the chapter.

45. I am grateful to John Ravenhill for drawing my attention to the potential contradiction between increased local sourcing and expanding Japanese component exports. See later in the chapter for a description of increased component imports by the Taiwanese electronics industry. For an earlier presentation of the evidence, see Mitchell Bernard and John Ravenhill, "Beyond Product Cycles and Flying Geese: Regionalization, Hierarchy, and the Industrialization of East Asia," *World Politics,* no. 47 (January 1995), pp. 171–209.

46 See D. Ernst, and D. O'Connor, *Competing in the Electronics Industry: The Experience of Newly Industrialising Economies* (Paris: OECD, Development Centre Studies), chapters 1 and 2.

47. Some important examples are Canon for compact copiers, laser and bubble-jet printers, and a new enabling technology for liquid crystal displays called ferro-electric LCD; Sony for Walkmans and camcorders; Sharp for active matrix-LCDs, a miniaturized and user-friendly camcorder based on Sharp mini liquid-crystal video displays, called VideoCam, and refrigerators based on "neural network" technology; and Fuji Photo Film's new 3.5-inch floppy disk, which is capable of recording fifty times as much data as those currently in use, opening up the possibility for floppy disks to dominate future markets for portable recording media for digital data.

48. Toshihiko Kinoshita, "Japan's Foreign Direct Investment in East and Southeast Asia: The Current Situation and Prospects for the Future," Export-Import Bank of Japan, Research Institute for International Investment and Development, Tokyo, 1994, p. 40.

49. World Bank, *Global Economic Prospects and Developing Countries* (Washington, 1995).

50. See the later subsection called "A Shift toward Decentralized Governance Structures."

51. Telephone interview, March 1996.

52. In terms of the market growth potential for such products, the orders of magnitude involved are mind-boggling. Since 1992 East Asia's demand for consumer electronics has grown on average by 15 to 20 percent annually. For China, demand growth

for consumer electronics was even higher and exceeded 20 percent a year. Market research in China (roughly 1.1 billion inhabitants) and Indonesia (more than 180 million inhabitants) shows that in these countries roughly 10 percent of the population can afford to buy a variety of consumer durables. It is thus hardly surprising that Japanese and Korean producers of consumer electronics, as well as Philips and Thomson, consider East Asia a strategic market.

53. For an analysis of the competitive strategies of Japanese and Korean electronics firms in Vietnam, see Dieter Ernst, "Developing Vietnam's Electronics and IT Industries: Observations and Suggestions," a report prepared for the Vietnamese government, July 1995.

54. Ernst and Guerrieri, "International Production Networks." This study analyzes how the spread of U.S. and Japanese production networks in Asia has shaped their trade links with this region in the electronics industry.

55. A very different pattern, however, prevails for China. The share of components in U.S. electronics exports to that country has actually declined, from more than 12 percent in 1990 to less than 2 percent in 1993, while electronic data processing and telecommunications equipment are by far the most important export categories. To a large degree, this stems from the successful market penetration strategies of AT&T and Motorola in the Chinese market for telecommunications equipment.

56. Computer penetration rates, as published by International Data Corp. The figures for Taiwan are courtesy of Institute for Information Industry, Market Intelligence Center, Taiwan. (Hereafter III-MIC.)

57. Taiwan, despite its flourishing PC industry, still has an unremarkable domestic computer market compared with that of the world's advanced nations; its computer penetration rate is now estimated to have reached roughly 77 per 1,000.

58. Figures provided by III-MIC.

59. For details, see Ernst, "Globalization, Convergence, and Diversity."

60. Miyashita and Russell, *Keiretsu,* p. 200.

61. Most of these Chinese suppliers are affiliates of Taiwanese computer companies. This is discussed later in the chapter.

62. Almost without exception, local computer markets in Asia are heavily protected and thus are difficult to penetrate. The only exceptions are Hong Kong and Singapore. Foreign companies are unlikely to succeed without powerful local joint venture partners that can guarantee access to distribution channels. Penetrating these markets normally requires three steps. The foreign computer company must first establish sales and marketing joint ventures, which, if they work, will be followed by consignment assembly by a local company. Only if these first two phases are successful will the foreign company consider investing in its own manufacturing affiliate. This gradual approach is meant to reduce the risk of international production and to balance the limited resources available with the need to match production sites and major potential growth markets.

63. See the excellent analysis of the dialectics of outsourcing and integration in the hard disk drive industry in Clayton M. Christensen, "The Drivers of Vertical Disintegration," Harvard Business School, October 1994.

64. Changes in the product composition of China's electronics exports to the United States clearly indicate this upgrading in capabilities. Traditionally, China's electronics exports have been dominated by two product groups: consumer electronics and house-

hold appliances. The combined export share of both products reached a peak of 80 percent in 1989 and fell to roughly 64 percent in 1993. Since 1990 the share of computer-related products has been increasing at a rapid pace: it doubled from less than 6 percent to nearly 12 percent between 1990 and 1992 and increased further to almost 15 percent in 1993. Ernst and Guerrieri, "International Production Networks."

65. For a detailed analysis of the sophisticated mass production capabilities of Korean *chaebol,* see Dieter Ernst, "What Are the Limits to the Korean Model? The Korean Electronics Industry under Pressure," research paper prepared for BRIE, 1994. An assessment of the strengths and weaknesses of the Taiwanese electronics industry can be found in Dieter Ernst, "New Opportunities and Challenges for Taiwan's Information Industry: Where Does It Stand Compared to Japan and Korea? And What Are Taiwan's Strategic Options?" study prepared for III-MIC and BRIE, 1997.

66. In a study for the Sloan Foundation project on the globalization in the electronics industry, I show that U.S. electronics firms were able to improve their specialization by relying on a combination of the following three strategies: (1) an early redeployment of final assembly and testing activities to a few locations in East Asia; (2) the outsourcing of an increasing variety of value-chain activities to Asian contractors, first in Japan, then in Korea and Taiwan, and now also in China and other Asian countries; and (3) a systematic rationalization of the international production networks that have emerged as a result of these activities. By redeploying lower-end stages of the value chain to Asia, U.S. electronics companies were able to concentrate on what they do best, namely, on product design and the definition of global brand names and architectural standards and on the control of distribution channels. A focus on such higher stages of the value chain generated high profit margins. In turn, that enabled U.S. companies to stay ahead through aggressive new product development strategies. See Ernst, "From Partial to Systemic Globalization."

67. Important contributions include David Mowery and Nathan Rosenberg, *Technology and the Pursuit of Economic Growth* (Cambridge University Press, 1990); Martin Kenney and Richard Florida, *Beyond Mass Production: The Japanese System and its Transfer to the U.S.* (Oxford University Press, 1993); Imai, "Patterns of Innovation and Entrepreneurship in Japan"; Kim Clark and Takahiro Fujimoto, *Product Development Performance: Strategy, Organization, and Management in the World Auto Industry* (Harvard Business School Press, 1991); Ikujiro Nonaka and Hirotaka Takeuchi, *The Knowledge Creating Company* (Oxford University Press, 1995); and Lewis M. Branscomb and Fumio Kodama, "Japanese Innovation Strategy: Technical Support for Business Visions," CSIA Occasional Paper 10 (Harvard University, Center for Science and International Affairs, 1993).

68. The following account is based on T. Abo, "Overseas R&D Activities of Japanese Companies" (in Japanese), *Journal of Science Policy and Research Management* (Tokyo), vol. 7, no. 2 (1992); and interviews in the Japanese electronics industry, November 1993 and June 1995.

69. Basic manufacturing support services are defined to include such activities as calibration and testing, die and tool services, (preventive) maintenance and repair, and quality control.

70. Based on interviews at Sharp headquarters, June 1995.

71. Shigeki Tejima, "Toward More Open Corporate Strategies: Will Japanese Firms Take Those Strategies to East Asia?" Export-Import Bank of Japan, Research Institute

for International Development and Development, Tokyo, May 1996, reporting on the most recent annual questionnaire survey of the Export-Import Bank of Japan.

72. Dennis Tachiki, "Extending the Human Capital Formation Process: Transnational Corporations and Post-Employment Training," *RIM: Pacific Business and Industries,* vol.1, no. 23 (1994), pp. 33–44.

73. Most of my interview partners in Japanese electronics firms emphasize that they cannot merely replicate the U.S. model of international production. Their view supports my argument that changes in the organization are path dependent, in the sense that they are shaped by peculiar features of the domestic production system. For Japanese firms this means that they have to come up with an organizational response that is based on and amplifies the strengths of their domestic production system. One prominent example is the distinctively different approach of Japanese electronics firms to human capital formation in Asia.

74. During the peak of the recession, when most Japanese electronics firms faced a serious profit squeeze, Japanese parent companies actually used sophisticated transfer-pricing techniques to transfer back home any profits made in Asia. Interview with Japanese venture capital firm, November 1993.

75. Based on *Business Times* (Malaysia), September 1995, p. 4.

76. Both NEC and Fujitsu are involved in some smaller projects but fall well behind the market leaders from the United States (AT&T and Motorola) and Europe (Ericsson, Alcatel, and Siemens).

77. In 1994 the share of U.S. computer companies of China's brand name desktop market was 42 percent, with Compaq (15 percent) as the dominant market leader. For notebook PCs the share of U.S. companies was 30 percent, and Taiwan's Twinhead was the largest non-U.S. brand vendor, with 5.5. percent. Toshiba, which worldwide is among the market leaders in laptop computers, had managed to acquire no more than a miserable 1 percent market share, even behind Philips's 2 percent share. Market share figures are courtesy of III-MIC.

78. Based on interview at Matsushita headquarters, November 1995.

79. Based on Henny Sender, "Be Prepared: China Is a Minefield for Even the Sturdiest Firms," *Far Eastern Economic Review,* April 27, 1995, pp. 61–62.

80. Consumer electronics is one of the areas in which tariff reduction has achieved considerable progress in the context of AFTA, the regional free trade zone of ASEAN countries. See Tan, chapter 4.

81. All these four countries have recently received a major inflow of investment and subcontracting arrangements by Japanese electronics companies, with China being by far the most important recipient. Vietnam has also succeeded in entering the Japanese production network: Fujitsu, in early 1996, established a printed circuit board assembly plant in Vietnam that could pave the way for further Japanese investments.

82. Cover story on ASEAN, *Far Eastern Economic Review,* August 11, 1994, p. 32.

83. Information provided by III-MIC.

84. Between 1992 and 1994 the overseas production of Taiwanese computer firms more than tripled, from about $970 million to more than $3 billion. And in value terms the ratio of overseas production out of Taiwan's total PC production increased from 10.4 percent in 1992, to 14.9 percent in 1993, 20.6 percent in 1994, and 27.2 percent in 1995. For 1996 the overseas production ratio is estimated to increase to 29.5 percent. Throughout this short period, annual growth in overseas production value was consis-

tently over 70 percent, which implies that overseas production today plays a critical role for the success and failure of Taiwan's PC industry. Ernst, "How Small Enterprises Internationalize."

85. Three additional factors were of equal importance: a highly motivated work force with a level of education that, in most other countries, is beyond the reach of SMEs; a highly flexible, hierarchical domestic supply base of loosely connected specialty producers that have enabled Taiwanese computer vendors to react incredibly fast to changes in market requirements and technology; and the role of government policies that initially facilitated market entry of SMEs and that now provide some of the externalities that are essential to upgrade this industry. For a detailed analysis, see Ernst, "How Small Enterprises Internationalize."

86. These and the following figures are courtesy of III-MIC, May–June 1995.

87. Ernst and Guerrieri, "International Production Networks."

88. TSMC, the world's leading silicon foundry, was able to capture an order volume of $20 million, primarily as a supplier to NEC and Fujitsu. Although TSMC's exports to Japan in 1994 made up only 0.5 percent of its annual sales, this share has increased to 2 percent in 1995, out of an estimated $1 billion in total sales. Other Taiwanese firms that were able to benefit include leading IC producers like Vanguard International Semiconductor Corp., United Microelectronics Corp., Hualon Microelectronics, Macronix International, Winbond Electronics, and Mosel-Vitelic, but also a number of much smaller companies like Tamarack Microelectronics, Everlight Electronics, Hi Sincerety Microelectronics Corp., Kingbright Electronics, and Mospec Semiconductor Corp.

89. These and the following trade figures are taken from Ernst and Guerrieri, "International Production Networks."

90. Figures provided by III-MIC.

91. The Japanese delegation was headed by the vice chairman of the Matsushita group and included a broad spectrum of semiconductor producers, users, and producers of other components, including NEC, Toshiba, Hitachi, Fujitsu, Mitsubishi, Matsushita, Oki, Sony, Sharp, Alps, Anritsu, Ricoh, Victor, Minolta, Sanyo, Seiko Epson, Yamaha, and Yokogawa.

92. "Japanese PC Makers Move Production In-House," EBN Online, July 1, 1996.

CHAPTER EIGHT

The China Circle and Technological Development in the Chinese Electronics Industry

Jean François Huchet

SINCE THE IMPLEMENTATION of the open door policy in China in 1978, the Chinese electronics industry has been one of China's fastest growing industries, to the point that China is now often regarded as the next superpower in electronics.[1] This expectation is normally based on a series of broad assumptions in which the impact of the China Circle holds a predominant position. China is supposed to share the same cultural background as its virtuous Asian neighbors and to belong to a vague but definite "Asian economic development model." In that way, China would be able to use the same development strategy as Japan and the Four Little Dragons (Singapore, Hong Kong, South Korea, and Taiwan) to build a strong electronics industry within its boundaries, combining a high export competitiveness with an import substitution policy for an enormous domestic market. Crucial to this strategy is China's ability to absorb a great deal of foreign technology to develop its technological capabilities. Here, too, the impact of the China Circle is supposed to be positive, allowing China to pick up some methods successfully adopted by its Asian neighbors and to benefit from foreign direct investment (FDI) from Japan and the Four Little Dragons.

Whether China is going to fulfill this prophecy remains to be seen.

Moreover, the effect of the China Circle on the Chinese electronics industry, especially on the accumulation of technological capabilities, is complex. Even if Japan's and the Four Little Dragons' influence on the Chinese electronics industry since 1978 has been important for production, investment, and export capabilities, the level of technological transfer has remained low, with China benefiting passively from the delocalization of the Japanese and the Dragons' production network in Asia. Both the Chinese national system of innovation and the globalization process in the electronics industry limit the significance that the technological accumulation experiments in Japan and the Dragons (which, it must be noted, are far from identical) can have. These also modify the way China will have to develop its technological capabilities before it can assume an active role in the China Circle electronics industry.

Major Trends in the Development of the Chinese Electronics Industry

In 1978 the Chinese electronics industry was mainly oriented toward military purposes. The total output in 1981 was 10 billion yuan, and the consumer electronics industry was virtually nonexistent, largely because of a political ban on consumer products.[2] Despite the Sino-Russian rift in 1960, the Chinese electronics industry was technologically and structurally (organization of research and development) heavily dependent on the Soviet Union. State and collective enterprises managed almost all total output. The production was historically concentrated in Shanghai and in Jiangsu province, but major investments were also carried forward during the Cultural Revolution in some inner provinces such as Sichuan, Guizhou, and Shaanxi because of the fear of foreign attacks on industrial bases in coastal provinces.

Since then the transformation of the Chinese electronics industry has been impressive. Electronics output grew from 10.0 billion yuan in 1981 to 186.5 billion yuan in 1994 (figure 8-1). In 1993, 25 percent of output was exported ($8.1 billion),* and China is now among the top ten suppliers of electronics products in the United States.[3] The Chinese electronics industry benefited from the boom in household appliances

*All dollar amounts are U.S. dollars unless otherwise indicated.

Figure 8-1. *Evolution of Chinese Electronics Output between 1981 and 1994*

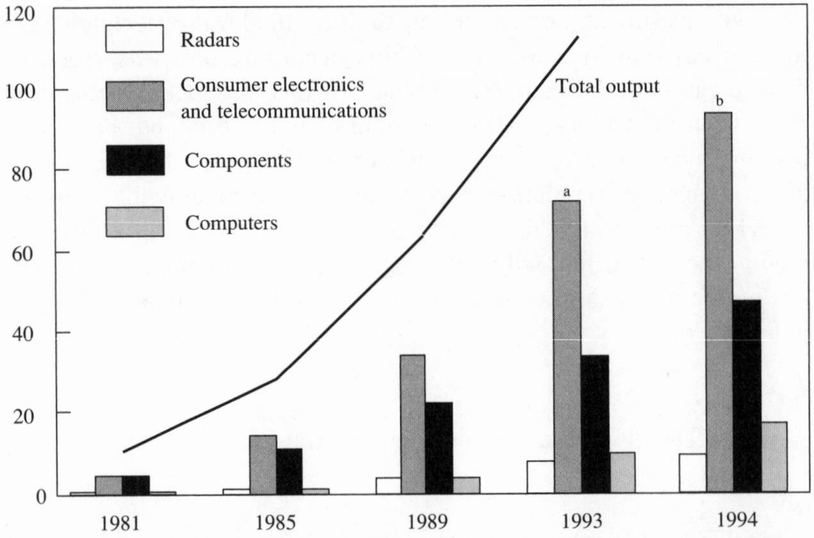

Source: *Zhongguo dianzi gongye nianjian* (China's electronics industry yearbook), 1994, 1990, 1987.

a. Of which consumer goods 60.1, telecommunications 12.

b. Of which consumer goods 78.2, telecommunications 15.

in the 1980s, especially in television sets and audio equipment. This boom in consumer electronics drove the industry's growth, not only by increasing production capabilities, with the import of more than 124 production lines, but also by boosting the consumption of television-related products, such as cathodic tubes and semiconductors. China has 140 television makers, which produced 13 million black-and-white and 13 million color televisions in 1993. In the same year seven of the ten biggest enterprises were devoted to consumer electronics production.[4]

The computer industry is progressively emerging with success stories by private ventures such as Sitong (Stone) in the 1980s and Lianxiang (Legend, a Sino–Hong Kong joint venture) in the 1990s. Lianxiang has been the second largest enterprise in electronics since 1993, although a big part of its turnover is based on the monopoly of commercial distribution of foreign brand computers in China. Foreign brand products dominate the domestic market: 70 percent of computer-related products sold in the Chinese Silicon Valley of Zhongguancun in

Beijing are foreign brand products. Imports represent 90 percent of software, 80 percent of printers, 80 percent of copy machines, 100 percent of facsimile machines, 90 percent of floppy disks, 100 percent of hard disks, 100 percent of microprocessors, and 50 percent of monitors.[5] With 25 percent of the market, Compaq is the leader in the sales of domestic PCs. But competition is growing stronger among low value-added products since the massive entry by Taiwanese and South Korean enterprises.

The component industry is dominated by passive components production, a field in which Chinese enterprises are progressively gaining international competitiveness. In 1994 exports of passive components totaled $2.1 billion, a 50 percent increase from 1993. But the situation in the highly strategic production of integrated circuits (ICs) is less promising. Chinese enterprises have not been able to catch up with the international pace of technological innovation. Only the Huajing Company in Wuxi, which benefited from huge investments from the Ministry of Electronics throughout the 1980s, is able to produce very large scale integration (VLSI) circuits matching international standards. The biggest enterprises in China are foreign joint ventures. Several attempts to build "national champions" in this industry failed during the 1980s, especially in Shanghai.[6]

With only 30 million lines, the Chinese telecommunications network remains one of the least developed in the world. The government is planning to invest 360 billion yuan to make the number of lines fifteen times greater.[7] Domestic producers are unable to match this huge technological demand. The Ministry of Post and Telecommunication (MPT) still holds a monopoly in the exploitation of the network, despite recent concessions given to the Ministry of Electronics. Only foreign companies are able to supply the required technology and negotiate their entry in this market through FDI. European companies operating in China hold the biggest market share in digital switchboard, with Alcatel Shanghai owning 40 percent of the market. In cordless telecommunications the situation is chaotic. The provinces often bypass the MPT's control of the market and negotiate directly with foreign companies without considering incompatible technological standards at the national level.

The regional distribution of production has changed, too, with the emergence of Guangdong province as the leader in the electronics industry. In 1982 Guangdong ranked sixth among regions, with an output of 549 million yuan, far behind the provinces of Shanghai and Jiangsu (2.2 billion and 1.6 billion yuan, respectively). But since 1993 Guangdong has become the largest producer of electronics products, with an output of 25 billion yuan, three times that of Shanghai. The provinces of Guangdong and Fujian are the main export bases in the country, exporting 38.5 percent and 32.4 percent of their total output, respectively, whereas most of the other provinces export less than 10 percent of their production (the sole exception is Zhejiang, with 23 percent).

The emergence of the provinces of Guangdong and Fujian as main production and export bases is directly linked to the delocalization of the electronics industries of Hong Kong (in Guangdong) and Taiwan (Fujian) during the 1980s. It is widely known that the reduction of the U.S. commercial deficit with Taiwan and Hong Kong and its growing deficit with China has occurred because of the wave of delocalization in China of Taiwanese and Hong Kong labor-intensive industries. A study on the influence of FDI in China shows that a total of 4,820 companies involving foreign capital were operating in the electronics industry in 1993, representing nearly one-third of total output (the figure was only 3 percent in 1985).[8] Data from the Ministry of Electronics indicate that foreign enterprises were responsible for nearly 55 percent of total exports in 1993 and 57 percent in 1994.[9] The foreign partners of 60 percent of the foreign-invested firms are from Hong Kong and Taiwan. Though their production constituted 9 percent of total output, they accounted for 25 percent of total exports in 1993.[10] In total, more than two-thirds of incoming electronics industry investment in China comes from Hong Kong, Taiwan, and Japan.

Delocalization and Technology Transfer from Japan, Taiwan, and Hong Kong since 1978

Foreign direct investment is by far the most important source of foreign technology in China (relative to licensing and complete plant and equipment purchase). Although for production, investment, labor, and exports, the impact of FDI is positive, the technological impact is still limited, with China benefiting passively from the globalization-regionalization (glocalization) process of Japan's and the Four Dragons' electronics industry in East Asia.

Japanese Delocalization

During the 1980s the technology transfer of Japanese companies in Chinese electronics was fairly limited. Except for a few cases of direct investment, such as Hitachi (in 1980 in Fuzhou, Fujian, to produce televisions) and Sanyo (in Shenzhen to produce tape recorders), which launched joint ventures, most Sino-Japanese technological and industrial cooperation took the form of equipment or license purchases.

In the early 1980s the Chinese government dropped its former technology transfer policy of complete plant purchases. With the four modernizations program launched in 1978, the official economic policy switched from extensive economic growth to intensive growth. On the technological side this policy was synonymous with reforming and modernizing existing production capabilities. State firms in the industry took advantage of a loose credit policy to import foreign technology massively, especially to produce consumer products like televisions, radios, tape recorders, and refrigerators. In a highly bureaucratic and provincially compartmentalized industrial system, most of these foreign equipment purchases could not develop a high level of economies of scale.[11] At the same time the level of software technology purchases (management methods, production, and quality control) compared with that of hardware was very small. In a highly bureaucratic environment, production grew, but most of the state enterprises could not accumulate technological capabilities, as had been done in South Korea, Hong Kong, Singapore, and Taiwan in the 1960s and 1970s. At the end of the 1980s both foreign studies and Chinese official statements acknowledged the limits of the policy governing technology imports in the electronics state sector.

Japanese firms in the electronics industry took advantage of this environment and sold at the local level the same production equipment to hundreds of small Chinese state firms, through the big and well-informed general trading companies (sogoshosha). (The most famous example of this remarkable business performance is the television industry.) Nearly 100 Chinese state firms imported 124 models of production equipment between 1978 and 1986. Japanese firms sold nearly 90 percent of this equipment, with limited technological assistance. This kind of short-term commercial objective with limited technology transfer fueled a negative image of the Japanese firms' involvement in the modernization of the Chinese industry.

After an initial period of enthusiasm immediately following the Sino-Japanese peace and the economic treaty in 1978, Japanese firms' involvement in FDI was cautious. Caution was especially evident after many Sino-Japanese cooperation contracts were canceled in 1981 by the Chinese (such as the highly publicized example of the Baoshan metallurgic complex in Shanghai). Uncertainties about the political situation as well as operating difficulties for joint ventures in China reinforced this cautious strategy.[12] The Japanese enterprises started to focus on technological and industrial cooperation with a limited number of Chinese firms, doing barter trade or operating on original equipment manufacturing (OEM) agreements, most of the time using Hong Kong expertise.[13] This cooperation allowed a bigger technology transfer to Chinese enterprises and reduced identification costs for Japanese enterprises in joint venture agreements at the beginning of the 1990s.

In 1991 Japanese enterprises began their fourth wave of delocalization (the other ones having taken place in 1960, 1970, and 1985). Since then Asia has constituted the main destination of Japanese FDI, with China the second most important Asian country for Japanese direct investment after Indonesia, up to 16.5 percent in 1993 from 9.8 percent in 1992. According to a recent Ministry of International Trade and Industry (MITI) survey, China will be the first destination for future Japanese direct investment.[14] Most large Japanese enterprises in the electronics industry recently launched big investment projects in China. The capital invested per project averages more than during the 1980s and ranks second only to that of the United States ($2.5 million per project). In electronics all the subsectors are now included, even ICs, which Japanese enterprises had been reluctant to delocalize in China. Some firms are now involved in many operations: Matsushita has sixteen production and distribution plants. Their subcontractors (first and second range in the *keiretsu* pyramidal organization) are following the same path. MITI is also ready to invest to support the creation of industrial parks (recently in Dalian and Hangzhou), which shows a Japanese long-term industrial commitment in China.

Since World War II Japanese firms have accumulated a huge in-house technological capability for production, investment, and innovation (according to the range and classification of technological capabilities used by development economists).[15] Even for standardized products or assembly activities, Japanese firms can transfer some of the technological capabilities required in China: production capabilities (quality control, efficient use of equipment, maintenance, repairs), design capabilities (for minor processes and product innovation), management, and marketing capabilities.

It is still difficult to reach definitive conclusions on Sino-Japanese operations in China, because most of them have just started or are still planning to start their production. For those that have been active for more than five years (Matsushita-Beijing in television electron tubes, Hitachi-Fuzhou in televisions, Mitsubishi-Shanghai in elevators, Epson-Shenzhen in printers, Sanyo in tape recorders and television sets, Mabuchi Motor–Dalian in tape recorder micromotors), case studies show how the transfer of elementary production and management capabilities took place.[16] Most of them invested heavily in human resources, which included training in Japan or on-the-job training done by Japanese mentors as well as the creation of professional schools and training centers. Furthermore, total quality control was implemented.

But the opportunities for technological learning are still confined to elementary capabilities of production in the workshop and in the intermediary management levels.[17] Transfer of investment and innovation capabilities is limited. The decisionmaking autonomy on technological issues is also limited. This confirms Michael Borrus's characterization (chapter 5) of Japanese technology transfer in Asia. The function of the Japanese transplant in China is not to innovate but to take advantage of low-cost production locations, with a subordinate role in the general strategy of Japanese industrial groups in East Asia, which are attempting to reinforce the coordination of their activities there. The plants focus mainly on assembly activities on standardized products, allowing Japanese firms to limit their technology transfer to elementary production and management skills. The plants end up adopting traditional hierarchical management features opposed to the innovative Japanese management system as described by several analysts.[18]

As for technological spillover in the local industrial system, the overall situation looks worse. Most of the Sino-Japanese joint ventures have faced big difficulties in finding reliable suppliers that could meet Japanese standards of quality, delivery, and price.[19] Here, too, the tendency of Japanese enterprises to work in a closed Japanese-only production network[20] can be applied to the Chinese situation. Nevertheless, Sino-foreign joint ventures are required by law to balance their hard currency expenses in yuan, which forces them either to increase their local supply or to export (which requires the products to be of good quality). Case studies show that different solutions have been found to overcome this problem.

The city of Dalian in the northeast, where nearly one-third of Japanese direct investment in China is concentrated, is working in cooperation with MITI and an investment company close to the Japan-China Association on Economy and Trade to build an industrial park. The purpose is to build good infrastructure and a training environment that will allow Japanese small and medium-size enterprises (SMEs) to invest in China as subcontractors for big Japanese affiliates or Sino-Japanese joint ventures. Similar projects have already been implemented in Hangzhou, in Zhejiang, and in Ningbo, Jiangsu, where Japanese investment is also significant. It is difficult to quantify the importance of technological spillover on Chinese local enterprises, but this solution recreates the Japanese *kereitsu* system in China and seems to leave little technological learning opportunities for local Chinese SMEs. Another solution, exemplified by Sino-Japanese joint ventures such as Fuji (Hitachi) in Fuzhou, is to organize local subcontractors in associations according to the *keiretsu* network pattern that exists in Japan. The subcontractors' association system became a good method for Fuji to work on price, quality, and delivery standards and to carry out transfers of production and management skills with its subcontractors.[21] This is the same type of relationship that exists among firms inside the *keiretsu* network created in Japan. The possible creation of a *keiretsu*-type industrial group in China was widely studied after 1978, but different experiments failed because of the bureaucratic environment in which Chinese firms still operate. It is difficult to know if this will become a common way for Japanese firms to operate in China, but because it represents far more than a firm-to-firm technology transfer, it could work as a powerful example of rationalizing interfirm relationships for Chinese electronics firms.

Delocalization by Taiwan and Hong Kong

As noted, Hong Kong and Taiwan are represented in 60 percent of the joint ventures operating in China in the electronics industry. Guangdong province has become the biggest producer in Chinese electronics because of the big wave of delocalization of electronics industries by Hong Kong. The same process occurred in Fujian province, with the delocalization of Taiwanese electronics industries. Fujian is currently the third biggest producer in China, surpassing Shanghai in 1993. As Chin Chung shows in chapter 6, the delocalization by Taiwanese electronics enterprises in China has been rapid since the beginning of the 1990s. Taiwanese operations in China account for 80 percent of China's total monitor output.

During the 1980s Hong Kong and Taiwan regarded China as a low-labor-cost reservoir for assembly activities to be reexported to the United States, Japan, or Europe, mostly in printed circuits, keyboards, monitors, telephone receivers, and video tapes. The average capital invested by a Hong Kong enterprise during the 1980s was one-third or one-fourth that of its U.S., Japanese, and European counterparts. Technology transfer was limited. Tehnological capabilities for innovation and investment were nearly nonexistent. Transfer of management capabilities was also minimal, with most of the joint ventures possessing only one foreign expert on their management staff. A 1990 study conducted by the Chinese–European Community Management Institute on the management practices in Chinese joint ventures revealed that management in Hong Kong firms was hierarchical, centralized, and authoritarian. Moreover, the training of employees was far more limited than in Japanese, U.S., and European joint ventures.[22] Local integration was also minimal, with most of the materials, semifinished products, parts, and components coming from the parent company or one of its affiliates. Still difficult to assess are so-called fake joint ventures (*jia yang-guizi*), in which no foreign contribution appears in the capital or in the management of the joint venture. Here the status of joint venture has been used by Chinese enterprises to benefit from a more flexible operating environment than the one available in the state sector (import-export, recruitment, firing, financing, management of foreign currencies, and labor policy). In this category, no technology transfer is possible.

China's Technological Emancipation and Glocalization of Electronics in Asia

Considering the national technological trajectory of China since 1949 and the globalization of the electronics industry, FDI is probably the better way for China to assimilate elementary technological skills, as against the state enterprises' strategy of licensing and purchasing plants and equipment. Nevertheless, FDI is not the only requirement for technological accumulation or for the emergence of a virtuous circle of meso-industrial development. At least four other interrelated factors affect China's ability to take full advantage of the globalization process of Japan's and the Four Dragons' electronics industries and to emerge as a regional active player in electronics. These are (1) China's position in the complex interdependent Asian production system; (2) China's ability to use access to its huge domestic market as a bargaining chip; (3) limitations on China's ability to assimilate FDI-related learning opportunities; and (4) limits to China's national system of innovation.

Asian Interdependencies

As described by William Cline, the traditional framework for electronics production in Asia models growing competition among countries after technological catch-up.[23] FDI is seen as the main transmitter of technology to latecomers, allowing them to compete with more advanced countries. The best example is South Korea, which is now challenging Japan as the world's leading producer of DRAMs. Korean firms spent more money in 1995 in semiconductor investment than did their Japanese counterparts ($2.8 billion as against $2.7 billion).[24] Many Japanese semiconductor firms worry about this growing competition and think that this technological catch-up may negatively affect them.

This traditional pattern raises the important question whether Japan and the Little Dragons can move toward higher value-added products. If they fail, the competitive pressures from China will be more important. They would then need to reduce technology transfer to China. Figure 8-2 shows that the Chinese continue to benefit from the complementarities between Asian countries in their electronics production, at least in the short term. But in the medium term, the structure of production may become more competitive.

It would be a mistake to generalize this model, however, since many factors tend to complicate the framework in which production takes place in a glocal industry. First, market volume is not stable: overseas markets in the United States and Europe are not the only ones to grow, since Asia (except Japan) and South America are consuming more and more electronics products. This growth allows entry of new competitors without entailing a reduced production capacity for the leaders.

Second, and more important, innovation continually tends to modify

Figure 8-2. *Comparative Share of Hardware Production in Asian Countries, 1992*

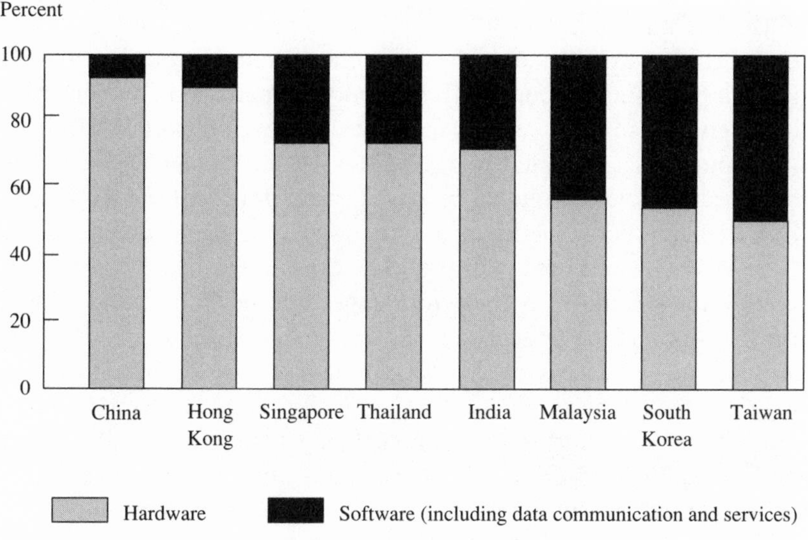

Source: Organization for Economic Cooperation and Development.

products' main characteristics. Generic products, such as telephones, computers, and machine tools, still exist, but they are continually being transformed by the effect of innovation. Innovation tends to create diversification of products, reinforced by flexibility in production processes, to account for customer requirements. This diversification creates new markets and allows the entry of new competitors.

Simultaneously, for every product one must consider both its content and the different activities that influence its competitiveness (R&D, design, assembly, distribution, and so on). Those two characteristics together can explain the growing regional networking process of production, one of the main elements of the glocalization process. Firms are not doing everything in the same territory. They use subcontracting and creation of affiliates, as well as new contractual arrangements (joint production, joint ventures, and OEM contracts) with other firms in different countries, to optimize their competitiveness. Firms tend to rely on networks for their production, which embraces different territories and different forms of cooperation. Because of the technological complexities, nations are becoming specialized in a specific activity in the value chain (R&D, design, production) and in certain kinds of products. Only the national official foreign commerce statistics give the illusion of a frontal competition among countries. It is true that China produces and exports complete electronics products, but in 1993, 66 percent of its total imports ($10 billion) were parts and components to be assembled in China and 91 percent of its total exports ($8 billion) were products assembled in China.[25] This shows that China has an intermediate role in regional production, but it also shows the inaccuracy of the traditional theoretical framework described by Cline.[26]

In this context, China will probably continue to benefit from technology transfer through FDI from Japanese and Four Little Dragon firms. A look at the Dragons' electronics industries indicates that they are entering a process of diversification and massive investments in higher value-added products. As Chung shows in her chapter, even if those industries' short-term investments in China remain in the assembly of standardized products, there are already some attempts to take advantage of the low cost of Chinese engineers to develop higher value-added activities. China attracts more and more investment from Hong Kong, Taiwan, and Japan in high value-added products or activities because of its low cost and big reservoir of qualified engineers willing to work for the higher salaries available in a joint venture instead of a public research institute or state enterprise.[27] Hong Kong firms like VTech (computers and mobile telephones), Varitronix (software), Legend group (computers), and Truly International Holding (LCD screens) are increasingly recruiting Chinese engineers in their mainland subsidiaries. Those companies are transferring more complex technological activities (such as design and creation of prototypes) to China.

At this point an important question arises: can this movement result in a powerful production system led by overseas Chinese (from Hong Kong, Taiwan, and Southeast Asia) that is able to compete with the U.S. and Japanese network production systems, as Michael Borrus suggests in chapter 5? There are two main factors that tend to limit such a possibility. First, the battle for access to the PRC domestic market has already begun, and it seems unlikely that an overseas-Chinese network production system could win against U.S., Japanese, or even European production systems. The Chinese government is well aware of the attraction of access to its internal market as a bargaining tool to get technology transfer from foreign companies and governments (as discussed in the next subsection). Even if the PRC enters the World Trade Organization, the recent industrial policies regarding automobiles, telecommunications, and ICs suggest at least two constraints on future foreign activity: it will be restricted to certain industrial segments in which domestic enterprises are not able to provide a good technological solution; and foreign companies will have to produce locally in cooperation with domestic companies and will be compelled to transfer their best technology. The market share of foreign companies in the PRC domestic market indicates a dominant position for the U.S., Japanese, and European firms. At present these are the only firms able to satisfy the Chinese leaders' industrial aspirations for technology transfer. Even if one takes into account the upgrading of the technological capabilities of the overseas-Chinese network production system that is currently under way, in the medium run it is unlikely to reach a dominant position.

Second, it is politically inconceivable that a powerful production system led by overseas Chinese and centered on the PRC for production can assign only low value-added activities to the PRC's territory. The PRC leaders have great economic aspirations. Generally speaking, the foreign presence is seen by the PRC leadership as a temporary way to reinforce domestic capabilities. Only when domestic firms have a dominant position in the production network will the situation be politically acceptable. This would depend on two important things: the PRC domestic firms, especially state enterprises, would need substantial technological modernization, which is still being hindered by many obstacles (as discussed later); and many important remaining political barriers would have to be removed. How could Taiwan, Singapore, and other East Asian nations willingly go against their own interests and share or give away technological leadership to PRC domestic firms in this China Circle production network? As Wang Gungwu has shown in a series of remarkable works, the overseas-Chinese reality is a complex one, formed by heterogeneous populations that have different historical backgrounds and belong to different nations. Those nations still have divergent national interests, which limits the possibility of a powerful and articulated overseas-Chinese production system in which the PRC domestic firms could hold technological leadership. Thus, realistically, the overseas-Chinese production system will continue to be subordinate to the U.S. and Japanese ones. This does not mean that it will not benefit from PRC production capabilities and the huge PRC domestic market but that it will do so in some hybrid form, as Ernst describes in chapter 7.

Japanese firms remain the principal and most important source of technology in Asia. Their delocalization in China seems irreversible, provided that political tensions do not interfere. The evolution of the investment environment in China is not the only explanation for the recent boom in Japanese direct investment. The *endaka*, or yen appreciation, which started in 1985, caused some structural transformations inside the Japanese industrial system. Most of the big Japanese enterprises reached the limit of productivity gains by management innovation and mechanization. The growth of labor costs in the most labor-intensive activities put pressure on the biggest firms to substitute foregn direct investment for their traditional links with SMEs inside the network of the *keiretsu*.[28] Anticipating this movement, or in response to it, many SMEs were forced to delocalize their production in order to survive, while developing high value-added production in Japan. China now is a major recipient of direct investment from Japanese SMEs.[29]

Furthermore, as Ernst mentions, there is growing pressure on Japanese transnational corporations in Asia to open up their production networks, especially because of the yen appreciation and the emancipation of the first-line subcontractors inside the *keiretsu*. The big question is how the Japanese can rely more on local subcontractors and open their production network without compromising the advantages of low-labor-cost production or losing competitiveness because of technology transfer. The increasing reliance on local subcontractors means a greater transfer of technological capabilities, which until now has been reduced to a minimum. Ernst insists that Japanese companies are increasing their R&D transfer capabilities in their Asian transplants. Even if no clear strategy is apparent, there is reason to believe that they will continue doing so. That requires product customization to cope with the emergence of a local market, better coordination with both local suppliers and Japanese implants, and the need to upgrade regional and local support service. As for the PRC, it is important to add one more factor to those listed by Ernst in chapter 7 to explain a possible evolution toward a greater opening of the Japanese production network. The pressure that the PRC central government has put on foreign companies to produce locally in cooperation with domestic companies, in exchange for access to the domestic market (which is becoming the most important investment motivation for foreign companies), might also accelerate the opening of the Japanese production network in China.

It is also important to note that exit costs are high for a company that has decided to invest in a foreign country. The parent company prefers to upgrade the technological level of its existing transplant rather than withdraw from the country. As Shujiro Urata has shown, Japanese producers of consumer electronics attempted to divide the entire production process into a number of subprocesses and transfer each to a country where that particular subprocess can be done most efficiently at the lowest cost.[30] The Four Little Dragons were assigned to be component suppliers. The Japanese transplants in Taiwan, which began in the 1960s, are still in operation even if the initial labor-cost advantage has shifted to China. The reason for continuing to invest and produce in Taiwan changed after the upgrading of the technological capabilities of the transplant. Considering the volume of Japanese FDI in China, it would be extremely costly for Japanese enterprises to withdraw even after labor-cost advantages narrow.

Nevertheless, the evolution of the role of the Japanese transplant in China will not be automatic. It is important that the host territory have the capability to adapt and evolve technologically to attract a higher value-added product investment. To feed its absolute advantages, the host country must upgrade its labor market, education and R&D systems, and its infrastructure. If China can do this, it can also benefit from the technological upgrading of Japanese transplants in the medium and long term.

The Domestic Market Bargaining Advantage

In the early stages of industrialization, strong, efficient bureaucratic control of FDI was widely used by Asian countries to integrate foreign technology into an overall technology development strategy.[31] More than other Asian countries, China has a natural bargaining advantage to attract technology from foreign firms, thanks to its potential domestic market. Although legislation to protect and attract FDI was adopted in 1978, China fully recognized its benefits only after 1986, when the accumulation of technological capabilities through licensing and equipment imports showed poor results in state enterprises. There then occurred a clear shift in the FDI strategy of the Chinese government.

Eager to attract FDI and technology, the Chinese government has enacted a series of new measures since 1986; it reinforced legislation on FDI, giving greater autonomy to the provinces to attract such investment, especially from Hong Kong, Taiwan, and overseas Chinese. China and Japan concluded an agreement in 1988 to grant protection to Japanese direct investment, but more important, Chinese authorities progressively eased the export rules applied to foreign-funded enterprises and joint ventures, allowing them to expand their sales on the national market. With labor costs increasing (at roughly 20 percent annually after 1988), the national market was becoming the main weapon used by Chinese authorities to attract FDI and technology. Estimates of cost savings for Taiwanese companies in the computer industry in China's operation, furnished by Chin Chung, vary from 8 percent in monitors to 10 percent in motherboards, 16 percent in switch power supply, 21 percent in keyboards, and 22 percent in mice. These cost savings are thin and can disappear altogether, since the Chinese macroeconomic environment includes high inflation and fixed exchange rates. Thus the size of the local market appears to be much more important, as confirmed by the different polls conducted on investment motivations in China for Japanese enterprises.[32] The Chinese domestic market in electronics expanded more than six times in eight years, from 28.6 billion yuan in 1985 to 175 billion yuan in 1993.[33]

China's technological acquisition strategy is clear. It allows foreign firms access to the domestic market in exchange for technology transfer through joint production or joint ventures. Considering the great potential of the domestic market, the authorities are using this bargaining advantage to make foreign firms compete with each other's technological bids and to push them to transfer the most recent technology and higher value-added activities. This strategy is now systematically integrated into the industrial policy for the development of those sectors considered highly strategic, such as automobiles, telecommunications, and integrated circuits.

In regulated industries such as telecommunication infrastructure, China possesses great bargaining advantage over foreign transnational corporations. Recent telecommunications deals with world leaders, such as AT&T on technology transfers of digital switches, microelectronics, network management, optical transmission products, cellular systems, systems integration, and research cooperation with Bell Laboratories, illustrate this capability. Japanese enterprises are also rushing to produce in China to avoid being left out of the game; Fujitsu and

NEC produce digital switches to erase their late-mover disadvantage in the Chinese market relative to European producers such as Alcatel Shanghai. The same logic applies in the recent $5.18 billion deal signed in March 1995 to upgrade telecommunications in the fast growing area of Shanghai, Hangzhou, and Nanjing, a deal that includes the major Japanese telecommunications companies and the biggest Japanese banks and service companies, helped by government agencies.[34] Part of the deal involves coproduction with Chinese enterprises. It demonstrates that Japanese enterprises, having been reluctant to transfer technology for a long time, are now being forced to do so to avoid being ousted by other global players.

Yet the ability to leverage domestic market access could be more limited in other products for which state control is weaker. Consumer electronics is probably the best example to illustrate the discrepancy between comprehensive industrial policy and its actual implementation. At the beginning of the 1980s the Ministry of Electronics wanted to limit the purchase of foreign technology to a few firms to create big industrial groups similar to those in Korea. The results were different from what was envisioned. There were 124 production lines imported by nearly the same number of television producers. This situation caused great structural difficulties in the television industry (financial difficulties, overcapacity, bad assimilation of imported technologies, lack of economies of scale, and so on).

There are many reasons for this lack of ability to implement industrial policy. An important factor is the conflict of authority within the state sector, in which the bureaucratic authority of the central government ministries overlaps with the regional authority of the provinces (*tiaotiao kuaikuai*). These structural problems plague industrial policy implementation in China. Taiwanese, Hong Kong, Singaporean, and Japanese firms are well aware of these difficulties. They know that obstacles to entering the Chinese market by the big door (through central ministries, according to their prices and technology requirements) can be overcome by using the little door (through provincial channels). In this context, the natural bargaining advantage of Chinese policymakers is greatly reduced.

Limits of FDI Strategy

Joint ventures are by far the most dynamic enterprises in terms of production, sales, benefits, exports, and productivity in China's electronics industry. State enterprises still lag far behind in competitiveness.[35] Yet the output of state enterprises still makes up a good portion of national output, 47 percent in 1994, according to the Ministry of Electronics.[36] As mentioned earlier, their technological modernization through technology imports (mainly hardware) was not very successful. Figure 8-3 shows how 1986 (the beginning of the seventh five-year plan) can be considered the turning point in Chinese technological acquisition policy, with the promotion of FDI as the main channel for acquiring technological capabilities.

Despite problems surrounding the control of economic activities by foreign interests, the central government determined that FDI can transfer more technological capabilities in the "soft" component of technologies (such as management, quality control, and marketing) than can be transferred by licensing or complete plant and equipment purchase, as demonstrated by the failure of the state enterprise strategy. In addition, the government adopted FDI policies to take advantage of the structural transformations inside the industrial systems of Japan and Asian newly industrializing economies. Asian NIEs (excluding South Korea, given the limited role of FDI there) offer a good example of how to benefit from FDI to accumulate technological capabilities in the first stage of industrial development.[37]

Nevertheless, the technological development of Asian countries also proves that FDI cannot substitute for indigenous technological effort for a country that aims at entering a virtuous circle of technological development. First of all, FDI in China is still mainly devoted to assembly operation and low-tech products. Even if the enterprises face high exit costs, they will not transfer higher value-added products or activities to their subsidiaries if they fail to find locally important inputs such as local subcontractors or infrastructure, human resources, a good education system, and access to financing and to public R&D. Economists consider the "adaptation capabilities of nations" to be their ability to transform their production system to attract higher value-added products and activities through FDI.[38] This capability is directly linked with the quality of the general environment in which firms are operating, including the education system, local subcontractors, and so on. Although this quality of the environment can be partly improved by the firm itself through technology transfer, training centers, and training of subcontractors, the main role obviously rests with the technological policies and technological efforts of the host country. The Four Little

Figure 8-3. *Evolution of Foreign Technology Purchases in China, 1983–92*

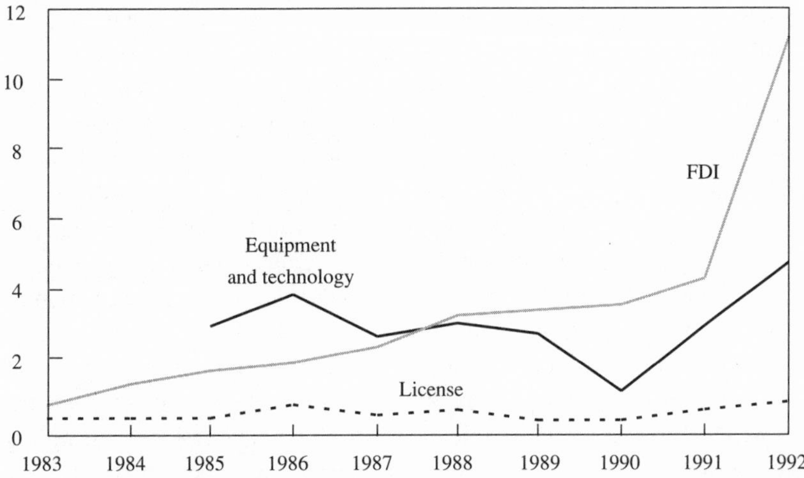

Source: Kang Songping, "90 niandai Zhongguo jishu yinjin de xin geju" (New situation for technology transfer in China in the 90's), in *Guanlishijie,* no. 1 (1994), pp. 169–71.

Dragons demonstrated a great capability to improve their education systems, to upgrade the technological capabilities of SMEs, to rationalize and develop public R&D, and to improve the financing system.[39]

Second, the growing networks of production started by glocalization in the Asian electronics industry can have some negative effects on the opportunities for technological learning by enterprises in developing countries. The possibility for one firm in a developing country to join the production network of a big transnational corporation constitutes a good opportunity to receive transfer of technology.[40] That is particularly true for elementary technological capabilities (production, equipment utilization, maintenance, repair, quality control), since the transnational corporation must impose certain quality standards in the production of parts and components by its affiliates or cooperating firm. But most of the equipment and technology are standardized by the parent company, and inputs come predominantly from abroad for the main purpose of export to other countries. This type of cooperation also applies to production agreements, such as OEM–value-added reseller agreements, under which local producers in developing countries are licensed by brand name companies to make their products under a strictly controlled set of specifications. According to Morris-Suzuki, technological autonomy can be significantly reduced for firms in developing countries that join the production network of a transnational corporation, because that limits the choice and adaptation of technology.[41] In this context, the modification of technology to suit local needs becomes very difficult. Moreover, property rights in joint ventures (and of course in wholly owned subsidiaries) leave the main power for technological decisions in the hands of the parent company. The firm assimilates elementary technological capabilities better and faster through its integration in the production network of a transnational corporation, but it could lose its technological autonomy to acquire higher levels of technology capabilities (design, incremental improvements, investments, innovation).

Third is a point directly related to the preceding one: the unpacking of technology constitutes a very important process that creates opportunities to learn new technological capabilities. Japanese technological development is a good example. From the Meiji period to the late 1970s, a large part of the technological activity of the biggest firms in electronics was precisely to understand, adapt, and import foreign technology. Mark Fruin, in his historical analysis of the Japanese enterprise system, shows how organizational structures and human and financial resources were focused on the adaptation of imported technology.[42] Japanese enterprises benefited not only from the traditional principle of "learning by doing" but also from what David J. Teece calls "learning by having learnt."[43] The process of trial and error used by Japanese firms in absorbing foreign technology can be regarded as an innovation process, and it is not surprising that the main strength of their current innovation is organizational. All that is to a great extent a consequence of Japan's historical experience in adapting foreign technology.[44] This also applies to a limited extent to South Korean and Taiwanese companies.[45] Chinese state enterprises lag far behind Japanese firms in financial resources devoted to adapting imported technology. A recent Chinese survey shows that for 1.0 yuan of imported technology Chinese state enterprises spend 0.7 yuan, as against the 10.0 yuan spent in Japanese enterprises.[46]

The general context in which industrialization takes place has changed a lot since Japan and the Dragons underwent their modernization. FDI is now one of the main features in the globalization of the electronics industry. For China, FDI provides the fastest access to elementary technological capabilities. Therefore, China will probably not go through the same experience as its Asian neighbors. However, this does not invalidate the fact that their experience demonstrates how FDI cannot substitute for a systematic indigenous technological effort in creating a virtuous circle of technological development.

China's National System of Innovation Deficiencies

Innovation incentives in Chinese electronics enterprises and indeed in the entire Chinese state sector are still very low.[47] A 1993 survey on innovation in fifty large and medium-size enterprises in Shanghai during the seventh five-year plan (1986–90) showed that only one-third of the enterprises are actively carrying forward a program of product development, with most new products remaining at the stage of sample or prototype.[48] Only 9 percent of these enterprises have quality standards that match international levels, and 80 percent are still operating on production equipment from the 1950s or 1960s. The big difference between enterprises under different ownership systems is maintained even in Guangdong, the most competitive province in China. A 1994 survey by the Research Bureau of the Communist Party for big state enterprises in Guangzhou found that only 16.6 percent of the 222 enterprises had modern equipment (corresponding to the international standard of the 1980s); most were still operating with 1940–60 equipment coming from Russia and had low quality and productivity levels (10 to 12 percent of the current world average level of productivity).[49]

The idea of national systems of innovation has been described by several authors.[50] Some critical issues about the environment in which Chinese indigenous technological efforts are taking place are worth noting. Internal factors such as poorly coordinated industrial policies, low innovation incentives for enterprises because of the bureaucratic environment, and low education investments severely limit the ability of the Chinese electronics industry to enter a virtuous circle of technological development.

Human Resources. It is widely recognized that the use of human resources in China is plagued with deficiencies. General problems with the education system and the role of intellectuals in Chinese society affect the motivation and the quality of human resources in China.

More directly connected with the industrial system is the continuing rigidity of the labor market, both in recruitment and transfer of employees.[51]

Many joint ventures face difficulties recruiting a good staff at the level of medium and high management.[52] Moreover, case studies indicate high levels of job hopping and staff turnover, which discourage foreign companies from investing more heavily in high-tech products.[53] It is important to note that the contribution of Japan and the Dragons to the development of human resources in high value-added products is still low and can play a critical role in China in the future. Public institutions and state enterprises will suffer from a flight of their best employees to foreign enterprises in China and throughout Asia. Japanese enterprises are importing 8,000 engineers from China annually on one-year contracts.[54] This is a big drain on the labor market of trained people for state enterprises, which are not attractive anymore for young trained people.[55]

The training problem that accompanies FDI should also be judged in the light of the potential spin-off of technological capabilities in the local industrial system, such as training of subcontractors, recruiting by Chinese firms of joint-venture or foreign-funded-enterprise employees, and creation of new firms.[56] The future depends on the evolution of the institutional environment, paticularly on whether it will give incentives both for creating high-tech start-up enterprises and for recruiting and attracting employment in Chinese domestic enterprises.

Industrial Policy. The central government must be able to implement an industrial policy. The problem is not to set objectives and elaborate a comprehensive plan. Central ministries are well endowed with efficient bureaucrats and well informed about the foreign situation. For example, comprehensive technological development plans such as the *Kejibaipishu* (White book on scientific development) or *Xinxi jishu fazhan zhengce* (Policy for development of information technologies) contain intelligent and up-to-date discussions of the technological situation.[57] Rather, the problem is the inefficiency of industrial policy for specific sectors, including the electronics industry. For instance, the development of a national production capability of video recorders failed. Chinese industrial policy in that sector began at the end of the 1970s, at nearly the same time as the more successful Korean experience.

The national production of integrated circuits is also currently facing many structural problems.[58] IC development was declared a priority by the central authorities at the beginning of the 1980s. However, it was only after 1986, with the publication of *Xinxijishu fazhan zhengce* (Policy for development of information technologies), edited by the Leading Group for the Revitalization of the Electronics Industry from the State Council, that the Ministry of Electronics tried to implement a comprehensive strategy. The main objective was to build a national production capability for integrated circuits within ten years and to contribute to the development of other segments of the electronics industry (mainly consumer electronics, telecommunications, and computers). The original program for the eighth five-year plan (1991–95) was to establish four national enterprises capable of producing more than 100 million ICs each and to become a medium-size producer on the world market by the year 2000. To realize this objective, central authorities sought to concentrate their financial aid on a small number of national enterprises in Jiangsu, Beijing, and Shanghai by a massive program of technology import. By the mid-1990s the objectives of the plan were far from being achieved. Some progress was made in meeting the technologically less demanding requirements for ICs for consumer electronics, but China still remains heavily dependent on imports to meet those needs. Domestic IC production covers only 20 percent of market needs, of which 53 percent is provided by foreign enterprises producing in China.[59] The biggest company in this sector is a Sino-foreign joint venture, the Belling Company, a subsidiary of the Sino-French telecommunications company Alcatel Shanghai. The leading domestic producer, Huajing, in Jiangsu province is probably the only Chinese producer capable of meeting international standards on standardized VLSI products (RAM products). Generally speaking, national producers are incapable of maintaining the pace of innovation required in this industry, which requires billions of dollars of investment in R&D and fixed assets for each generation of products.

That is why China is so eager to promote industrial and technological cooperation between national producers and the big global players. Central authorities want a few national producers capable of maintaining the pace of innovation but do not want to allow foreign companies to entirely dominate the domestic market. The results of this policy remain to be seen, especially since global players are not given clear standards that define their contributions and benefits. The problem of inconsistent national policy is clearly illustrated in other sectors such as telecommunications, as in the conflict between the Ministry of Post and Telecommunications (MPT) and the commercial arm of the Ministry of Electronics, Unicom, over the long-distance and cellular phone markets. The MPT refuses to allow foreign service companies entry into this market, and until very recently it also denied its rival, Unicom, access to its infrastructure. This prevented Unicom from developing a sufficient market share and endangered the active program of collaboration with foreign companies that Unicom had initiated.[60] Many foreign companies complain about the discrepancies between the formal rules of the MPT monopoly and the actual behavior of nearly autonomous local phone companies. For example, by allowing dozens of networks of cellular phone and paging operations to exist, the Chinese government has inevitably promoted chaos.[61]

To upgrade the efficiency of its industrial policy in the electronics industry, the central government recently combined sixty-two state-run electronics firms into a conglomerate, the China Electronics Corporation, which will focus on microelectronics and telecommunications equipment. This appears to be just another attempt by the Ministry of Electronics to reorganize the industry and the relationship between the bureaucracy and the state enterprises.[62] It is unlikely that simply forming a large conglomerate will improve that relationship; on the contrary, by reducing the degree of competition in the industry, such a move may be a step backward. More generally, since property rights and the problems created by a highly bureaucratic operating environment in state firms have not been solved, it seems difficult for the Chinese government to improve the efficiency of its industrial policy.[63]

Industrial Structure. The electronics industry in China also suffers from a lack of economies of scale and from the high level of self-sufficiency of individual enterprises (*daerquan xiaoerquan*). There were nearly 18,000 enterprises officially operating in the electronics industry in 1993. But in 1992, out of the 500 largest Chinese enterprises, only 22 were operating in the electronics industry, down from 31 in 1988.[64] The 22 large enterprises in the industry achieve a turnover of more than 1 billion yuan; the biggest enterprises attain a turnover of 3.5 billion yuan, a sum that is still very low compared with the main transnational corporations.[65] Matsushita, Hitachi, and Sony together produce more color television sets than all the 93 Chinese color television producers combined. This problem plagues the entire electronics industry but is particularly acute in semiconductors, audio products, personal computers, and optical fibers.[66]

Most explanations for this situation point to structural deficiencies in the Chinese industrial system and to its provincial and ministerial compartmentalization, as well as to the poor management of technology imports during the 1980s.[67] These realities constitute the major obstacle keeping state enterprises from entering into a virtuous circle of technological development.[68]

Financial Resources of the Enterprises for Innovation. R&D expenses in China are still far behind the levels of other industrializing countries, at only 0.7 percent of GNP.[69] In the highly strategic microelectronics industry, state enterprises spent an average of 0.6 percent of their sales in R&D activities during the 1980s, far below international levels.[70] There are many explanations for this phenomenon. Between 1980 and 1992 the average profitability in the Chinese electronics industry dropped nearly 25 percentage points.[71] In addition, given their responsibilities for social services (the *danwei* system) and the aversion to structural risk in the state sector, enterprises did not follow the official policy of spending 50 percent of their retained profits (*liucheng*) on technological modernization activities. During the 1980s state enterprises spent more money on capacity investments, employee bonuses, and fringe benefits.[72]

The capital depreciation rate and public funds are still too low to compensate for these negative factors. The official rate of depreciation in the industrial sector is 5.3 percent, but in reality enterprises still have to pay taxes on depreciation funds, which can be up to 50 percent.[73] Public funds to help enterprises in R&D activities are too low and spread among too many ministerial entities to be effective.[74] In a survey by Jiang Qingyun and Shi Lin, most of the enterprises responded that the main policies were not effective because of the strong bureaucratic intervention in their management.[75] The enterprises, which would prefer to operate with the same freedom enjoyed by joint ventures, blame state sector regulations for the lack of incentives to innovate. State enterprises increasingly find it impossible to compete with joint ventures. The annual grading of enterprises operating in the Chinese electronics industry (in *China's Electronics Industry Yearbook*) shows that the rising stars in many electronics subsectors are Beijing-Matsushita, which has a leadership position in electron tubes; Shanghai

Belling, predominant in integrated circuits; Fuji television (Hitachi), Sanyo, and Philips for television; Alcatel Shanghai, Ericsson, Siemens in telecommunications; and Legend in computers. The list does not include any Chinese state enterprises.

Conclusion

The Chinese electronics industry has experienced deep transformations since 1978: steady growth in production and exports, the diversification of production, and the emergence of a civil industry in consumer electronics. Since the end of the 1980s China has also benefited from the delocalization of the Japanese and the Four Little Dragons' electronics industries, which represent the major part of FDI in China. Although FDI played an important role in attracting huge amounts of capital, equipment, and new export networks, it is still concentrated in low value-added products and provides minimal transfer of technology. Even with Japanese enterprises, the opportunities for technological learning in joint ventures are still confined to elementary capacities of production at the workshop and intermediate management levels. Transfer of investment and innovation capabilities are very limited.

China is in a good position to continue to benefit from delocalization, for the internal transformations of the Dragons' and the Japanese production system are pushing enterprises toward a globalization of their activities in Asia. In certain subsectors, in which central government control is still strong, China can benefit from a natural bargaining advantage with foreign companies (because of its potential domestic market) to attract higher value-added products and activities.

Nevertheless, the more important question is not whether technology transfer will occur but whether China can take advantage of this big transfer of production within its boundaries and become an active player in the electronics production network in Asia. China's indigenous development of its electronics industry has been limited by various problems, especially in the state sector. China's national system of innovation is still plagued by many structural deficiencies. China cannot expect to overcome those problems by relying only on FDI. An analysis of Japanese and the Dragons' historical technological development shows that they made an enormous, comprehensive, and systematic indigenous effort before entering a virtuous circle of technological development, in which FDI was just one element contributing to the process. Given a stimulating and favorable environment coming from the networking processes in East Asia of the Japanese and the Dragons' enterprises, China is probably its own greatest foe in hindering the emergence of its virtous circle of technological development.

Notes

1. Electronics comprises components, consumer electronics, telecommunications, and computers.

2. *Zhongguo dianzi gongye nianjian* (China's electronics industry yearbook), 1986 (Beijing: Dianzi Gongye, 1987), pp. 1–40. Hereafter *China's Electronics Industry Yearbook.*

3. Electronic Industries Association, *1992 Electronic Market Data Book* (Department of Commerce, 1992).

4. Enterprise Evaluation Center, "1988–1992 nian zhongguo daqiye fazhan de yiban qushi" (Common trend in the development of Chinese big enterprises between 1988 and1992), *Guanli Shijie* (Management world), no. 6 (1993), pp. 98–114.

5. Ding Gouping, Xu Heping, and Xie Xiaoxia, "Guoji jisuanji shichang ji dui zhongguo shichang de yingxiang" (Impact of the international computer market on the Chinese computer market), *Zhonguo Gongye Jingji Yanjiu* (China industrial economic studies), no. 5 (1994), pp. 29–32.

6. Jean François Huchet, "Transferts internationaux de technologie et industrialisation tardive: le cas de l'industrie électronique en République Populaire de Chine, 1978–1991," Ph.D. dissertation, Université de Rennes, 1993.

7. *Nihon Keizai Shimbun,* October 24, 1994.

8. *Guanli Shijie,* no. 3 (May 1996), pp. 79–90.

9. *China's Electronics Industry Yearbook, 1994,* p. II–10; *1995,* p. 18.

10. *China's Electronics Industry Yearbook, 1994,* p. II–10.

11. Huchet, "Transferts internationaux de technologie."

12. Akira Iriye, "Chinese-Japanese Relations, 1945–1990," in *China Quarterly,* no. 124 (December 1990), pp. 624–38.

13. Hideo Ueno, "Japan-China Processing Contracts and Compensation Trade," *China Newsletter,* no. 39 (July–August, 1982), pp. 2–9. (Japan External Trade Organization.)

14. *China Newsletter,* no. 109 (March–April 1994).

15. See Carl J. Dalhman, Bruce Ross-Larson, and Larry E. Westphal, "Managing Technological Development: Lessons from the Newly Industrializing Countries," World Development, vol. 15, no. 6 (1987), pp. 759–77.; and Sanjay Lall, *Promouvoir la compétitivité industrielle dans les pays en développement* (Paris: Organization for Economic Cooperation and Development [OECD],1990).

16. Jian-An Chen, "Japanese Firms with Direct Investments in China and their Local

Management," in Shojiro Tokunaga, ed., *Japan's Foreign Investment and Asian Economic Interdependence* (University of Tokyo Press, 1992).

17. J. F. Huchet, "Les transferts de technologies des entreprises japonaises dans l'industrie électronique chinoise," *Ebisu,* no. 8 (January–March 1995) (Tokyo: Maison Franco-Japonaise).

18. Masahiko Aoki and Ronald Dore, eds., *The Japanese Firm: The Sources of Competitive Strength* (Oxford University Press, 1994); Mark Fruin, *The Japanese Enterprise System* (Oxford: Clarendon Press, 1994); and Kazuo Koike and Takenori Inoki, eds., *Skill Formation in Japan and Southeast Asia* (University of Tokyo Press, 1990).

19. Jixun Zhang, "Problems in Direct Investment in China," *China Newsletter,* no. 103 (March–April 1993), pp. 13–21.

20. As described by Dieter Ernst in chapter 7.

21. Sino-Japanese joint ventures, like their Sino-foreign counterparts, targeted the Chinese enterprises inside the military industrial complex to develop their local supply. Such enterprises appeared to be much more technologically reliable than the civilian SMEs. It also gave opportunity to the military firms to consolidate their civilian reorientation and to assimilate new management capabilities.

22. China–European Community Management Institute (CEMI) and China Enterprise Management Association (CEMA), "The Management of Equity Joint Ventures in China," Beijing, 1990.

23. William R. Cline, "Can the East-Asian Model of Development Be Generalized?" *World Development,* vol. 10 (February 1982), pp. 81–90.

24. *Asian Wall Street Journal,* March 15, 1995.

25. *China's Electronics Industry Yearbook,* 1994, p. VI–3.

26. Cline, "Can the East-Asian Model of Development Be Generalized?"

27. IBM's International Software Development joint venture in Shenzhen is paying up to $1,000 monthly to its software programmers and is recruiting directly from the biggest Chinese universities.

28. Y. Lecler, *Partenariat industriel: La réference japonaise* (Limonest: L'Interdisciplinaire, Collection Technologies, 1993).

29. Ministry of International Trade and Industry (MITI), *Small Business in Japan,* 1993, White Paper on Small and Medium Enterprises in Japan (Tokyo, 1993).

30. Shujiro Urata, "Emerging Patterns of Production and Foreign Trade in Electronics Products in East-Asia: An Examination of a Role Played by Foreign Direct Investment," paper for Conference on East Asian Economic Development, MITI, Tokyo, April 6, 1995.

31. Pierre Judet, "Transfert de technologie: expériences dans les PVD et succès Asiatiques," *Revue Tiers Monde,* vol. 30 (October–December 1989), pp. 777–96; J. L. Enso, "The Choice of Technique versus the Choice of Beneficiary: What the Third World Chooses," in Frances Stewart and Jeffrey James, eds., *The Economics of New Technology and Developing Countries* (London: Frances Pinter, 1982); and Tessa Morris-Suzuki, *The Technological Transformation of Japan: From the Seventeenth to the Twenty-First Century* (Cambridge University Press, 1994).

32. Chen, "Japanese Firms with Direct Investments"; and Zhang, "Problems in Direct Investment in China."

33. *China's Electronics Industry Yearbook, 1994.*

34. *Reuters,* March 20, 1995.

35. See the annual grading of enterprises operating in the Chinese electronics industry in *China's Electronics Industry Yearbook.*

36. *China's Electronics Industry Yearbook, 1995,* p. 18. The statistics from the Ministry of Electronics are not entirely consistent with those reported by the State Statistical Bureau and cited by Naughton in the Introduction to this volume. The discrepancies, which are not large, may be explained by different definitions of what types of output are included in the electronics industry. The definition used by the ministry may be influenced both by areas of administrative jurisdiction and by judgments about commodity classifications.

37. Chi Schive, *The Foreign Factor: The Multinational and Republic of China* (Taiwan) (Hoover Press, 1990).

38. Georges Benko and Alain Lipietz, eds., *Les régions qui gagnent districts et réseaux : les nouveaux paradigmes de la géographie économique* (Paris: Presses Universitaires de France, 1992); and François Chesnais, "Science, technologie et compétitivité," *Revue S.T.I,* no. 1 (Autumn 1986), pp. 97–148.

39. Dalhman, Ross-Larson, and Westphal, "Managing Technological Development," pp. 759–77; and Lall, *Promouvoir la compétitivité industrielle dans les pays en développement.*

40. It can be made by multiple existing cooperation agreements, such as joint venture, joint production, subcontracting, and compensation trade.

41. Tessa Morris-Suzuki, " Japanese Technology and the New International Division of Knowledge in Asia," in Shojiro Tokunaga, ed., *Japan's Foreign Investment and Asian Economic Interdependence: Production, Trade, and Financial Systems* (University of Tokyo Press, 1992).

42. Mark Fruin, *The Japanese Enterprise System: Competitive Strategies and Cooperative Structures* (Oxford: Clarendon Press, 1994).

43. David J. Teece, "Economies of Scope and the Scope of the Enterprises," *Journal of Economic Behaviour and Organization,* vol. 1 (September 1980), pp. 223–47.

44. Morris-Suzuki, *Technological Transformation of Japan;* and Fruin, *Japanese Enterprise System.*

45. Judet, "Transfert de technologie"; Dalhman, Ross-Larson, and Westphal, "Managing Technological Development"; and Martin Bell, "Learning and the Accumulation of Industrial Technological Capacity in Developing Countries," in Martin Fransman and Kenneth King. eds., *Technological Capability in the Third World* (London: Macmillan, 1984).

46. *Keji Ribao,* June 12, 1991.

47. Richard Conroy, *L'évolution technologique en Chine* (Paris: OECD, 1992); and Denis F. Simon and Merle Goldman, eds., *Science and Technology in Post-Mao China* (Harvard University Press, 1989).

48. Jiang Qingyun and Shi Lin, "Jishu chuangxin: guoyou dazhongxing qiye de xianzhuan yu wenti" (Technological innovation: situation and problem in big and medium state enterprises), *Zhongwai Guanli Daobao,* no. 3 (1993), pp. 4–11.

49. Yi Zhifeng, Wu Yuezheng, and Wu Yuefeng, "Cujing Guangzhoushi dazhongxing qiye jishu jingbu duice" (Policy for technological progress in Guangzhou big and medium enterprises), *Keji Guanli Yanjiu,* no. 2 (1994), pp. 39–46.

50. B. A. Lundevall, *National Systems of Innovation* (London: Frances Pinter, 1992); Chris Freeman, *Technology Policy and Economic Performance: Lesson from Japan*

(London: Frances Pinter, 1987); and Chesnais, "Science, technologie et compétitivité."

51. M. Korzec, *Labour and the Failure of Reform in China* (St. Martin's Press, 1992).

52. CEMI and CEMA, "Management of Equity Joint Ventures in China"; and Chen, "Japanese Firms with Direct Investments."

53. Huchet, "Transferts internationaux de technologie"; and Zhang, "Problems in Direct Investment in China,"

54. *Far Eastern Economic Review,* September 23, 1993, p. 77.

55. The situation is probably going to deteriorate, given the current problems inside the state sector.

56. Schive, *Foreign Factor.*

57. Guojia Kewei Chubanshi, *Xinxi jishu fazhan zhengce, zhongguo kexue jishu lanpi shu di si hao* (Policy for development of information technologies) (Beijing, 1990).

58. D. Rehn and D. F. Simon, *Technological Innovation in China: The Case of the Shanghai Semiconductor Industry* (Cambridge, Mass.: Ballinger, 1988).

59. *China's Electronics Industry Yearbook, 1994,* pp. III–58.

60. Henry Sender, "Ambiguous Access," *Far Eastern Economic Review,* January 9, 1997, p. 78.

61. *Business Week,* November 8, 1993.

62. For a historical perspective on the reorganization of the relationship between bureaucracy and firms inside the electronics industry, see Huchet, "Transferts internationaux de technologie."

63. Hong Kong can become a huge financial pump for the privatization of the Chinese electronics companies through a listing in the Hong Kong stock exchange. The evolution depends on the different institutional reforms in China (banking, property rights, commercial laws). The Tokyo financial authorities recently refused to list Chinese companies on the Tokyo stock exchange because of their lack of rigorous accounting standards.

64. Enterprise Evaluation Center, "Common trend."

65. *Zhongguo Dianzibao* (China's electronics journal), June 15, 1994.

66. Chang Weimin, "Woguo gongye qiye guimo jiegou de shizheng fenxi" (Scale economies analysis in the Chinese industry), *Zhongguo Gongye Jingji Yanjiu,* no. 5 (1991), pp. 43–48.

67. Huchet, "Transferts internationaux de technologie."

68. Returns on investments are too low to be reinvested in technological development activities. These comprise new investments and innovation that can guarantee a stable pace for technological capabilities accumulation. See Dalhman, Ross-Larson, and Westphal, "Managing Technological Development."

69. Jiang Qingyun and Shi Lin, "Jishu chuangxin." In 1989 Britain spent 2.2 percent of its GNP, Japan and Germany 2.9 percent, and the United States 2.7 percent.

70. Japan major electronics producers spent an average of 7 percent of their sales at the end of the 1980s. Aoki and Dore, *Japanese Firm,* p. 184.

71. Zheng Yuxin, "Zhengque yiren woguo gongye jingji xiaoyi de zongti qushi" (Evolution of the profitability in the Chinese industry), *Zhongguo Gongye Jingji Yanjiu,* no. 6 (1994), pp. 52–58.

72. Geng Xiao, "Managerial Autonomy, Fringe Benefits, and Ownership Structure: A Comparative Study of Chinese State and Collective Enterprises," paper prepared for the World Bank, Washington, 1991.

73. Huchet, "Transferts internationaux de technologie."

74. Guojia Kewei Chubanshi, *Xinxi jishu fazhan zhengce, zhongguo kexue jishu lanpi shu di si hao* (Policy for development of information technologies) (Beijing, 1990), pp. 70–71.

75. Jiang Qingyun and Shi Lin, "Jishu chuangxin."

CONCLUSION

CHAPTER NINE

The Future of the China Circle

Barry Naughton

THE THREE ECONOMIES of Hong Kong, Taiwan, and the People's Republic of China, working in concert, have enormous potential. Despite the dramatic growth of output and trade from the China Circle in the years since 1985, these economies are still far from fully exploiting that potential. There are substantial opportunities to improve productivity and profit from the transfer of technology and commercial skills to the China mainland. Moreover, although obstacles to future growth exist, none are so fundamental that they could not be either circumvented by alert individuals or eliminated by reasonably effective policymaking. Thus all three political jurisdictions appear poised for a prolonged period of fast growth and dramatic structural change.

At the same time, the years 1996–97 are emerging as a clear watershed in the evolution of the China Circle. Both on economic and political grounds, fundamental changes in the external environment are bringing to an end one phase of development. The transition to another phase inevitably brings heightened uncertainty. Indeed, only one aspect of future economic development can be known with near certainty: the growth of the mainland Chinese economy will cause future development to be less focused on the "second China circle" (that encompassing Hong Kong, Taiwan, and the two PRC provinces of Guangdong and Fujian). Following extended growth, the China Circle will merge into a broader and more integrated East Asia, involving more of the PRC mainland economy.

However, the particular form in which integration occurs will depend critically on the policies adopted in Beijing over the next few years. It is fairly easy to project that economic forces will lead to a further accumulation of manufacturing capabilities, that rapid growth will continue, and that the existing China Circle will become more integrated into a larger mainland Chinese economy and a larger East Asian economic region. But substantial political risk exists because of the unpredictability of the government policy in Beijing, and to a lesser extent that in Taipei. The overall political risk comprises two kinds of uncertainty: the first relates to the possibility of conflict and other political problems between the PRC government and those of Hong Kong and Taiwan; the second relates to policy choices that will channel economic development into different forms. Of these, the most important is the continued uncertainty over how much external liberalization the Beijing government will allow.

The End of an Era?

The pattern of economic activity in the China Circle that exists today initially evolved in response to the specific opportunities and limitations present in the 1980s. Most important was the pressing need felt by businesses in Hong Kong and Taiwan to restructure their export operations, combined with the accommodative Coastal Development Strategy, which allowed those businesses to invest in the PRC and then move commodities in and out of the PRC with a minimum of taxation or administrative interference. The result was a pattern of partial economic integration that allowed for cross-border networks in export production but did not fully integrate those networks into the economy of the People's Republic of China. After 1993 that basic dynamic was augmented by new opportunities to access the Chinese domestic market through direct investment, and Hong Kong and Taiwanese firms were quick to respond to new opportunities. For a period, from 1993 through 1995, China seemed able to enjoy the best of both policies. Rapid export growth continued unabated, while foreign investment directed at the domestic market flooded in.

Ultimately, though, contradictions between these two strands of policy (one, export-oriented; the other, import substitution) were bound to emerge that would diminish the effectiveness of each. Since 1996 these contradictions have indeed emerged, and the end of the first period of China Circle formation is clearly upon us. This is most apparent in the changing rate of export growth. PRC exports grew at an annual average rate of 17 percent from 1984 through 1995. In 1996 export growth dropped to 1.5 percent, and growth is projected at only 2.3 percent for 1997. Clearly, this dramatic slowing in export growth is partly the result of short-term and one-time factors. Problems with the PRC's program of tax rebates for exports led to a significant drop in exports in the first half of 1996, while changes in the way the United States allocates textile import quotas disadvantaged China Circle exporters. The political and military frictions between the PRC and Taiwan caused a drop in cross-Strait trade in the first third of the year, though trade subsequently rebounded. These short-run factors produced a sharp deceleration in export growth in 1996 without necessarily depressing future growth prospects. Yet the 1996–97 slowdown also clearly signals the end of one era and the beginning of another; it marks the transition from one policy environment to a new one.

The new economic relations are clearly reflected in currency values. All the authors in this book have stressed the importance of Asian currency realignment as a catalytic force providing impetus to export restructuring, and thus to the formation of the China Circle. The Chinese currency depreciated dramatically after the mid-1980s, culminating in the devaluation and unification of exchange rates on January 1, 1994. At that point China's position was highly favorable. While the domestic market was still large enough and growing fast enough to attract domestic market–oriented FDI, costs for exporters—especially labor costs—were still extremely low. However, between the beginning of 1994 and early 1997, the Chinese currency appreciated almost 40 percent in real terms relative to the U.S. dollar.[1] As discussed below, the primary reason for that appreciation was the combination of massive inflows of foreign capital and domestically generated inflation. During the same period the U.S. dollar strengthened against other currencies, especially against the Japanese yen, which fell from about 90 to the dollar during 1994 to only 125 to the dollar in April 1997. As a result, the Chinese *renminbi* appreciated in real terms more than 60 percent against the Japanese yen. Needless to say, these changes in relative cost conditions have substantially reduced the attractiveness of China as an alternative export platform, particularly for the Japanese. Indeed, Japanese contracted investment in the northern coastal city of

Dalian, a favorite for export-oriented Japanese producers, dropped by almost 40 percent in 1996, as potential investors responded to increasing costs and shifted to other locations more favorable for servicing the increasingly attractive Chinese domestic market.[2]

There is also evidence that the relocation of China Circle labor-intensive manufacturing to the China mainland is winding down. There simply are not that many low-tech, labor-intensive manufacturers left in Taiwan or Hong Kong—almost all have moved to the China mainland. PRC exports of garments and miscellaneous manufactures grew to a remarkable total of $38 billion in 1995 from only $8.4 billion in 1989. During 1996, however, such exports increased by only 3.2 percent, approximately the same pace as Chinese exports overall. Indeed, one of the few real growth sectors for Chinese exports in 1996 was computer products, exports of which increased 40 percent, to $6.7 billion. In fact, the increase in computer-related exports accounted for nearly all the total increase in exports in 1996.[3]

Of course, these events do not signal the exhaustion of China's export potential, nor the end of the Chinese economic boom. But they do indicate that the exceptionally favorable circumstances enjoyed by the Chinese economy during 1994–95 have now receded somewhat. The overwhelming cost advantages that China possessed for a few years have at least temporarily disappeared. Evidence of this is clearly seen in the competitive relationship between China and ASEAN that Tan Kong Yam discusses in chapter 4. Although investment inflows into China remained strong in 1995–96, ASEAN maintained a position of rough parity, with investment approvals in Indonesia, Malaysia, and Thailand together running at 65–70 percent of Chinese contracted investment (see table 4-1). Thus the Southeast Asian nations have managed to maintain a competitive position relative to China notwithstanding the enormous potential attractions of the Chinese market.

China still has immense comparative advantage in the production and export of labor-intensive manufactures. And because of its huge and increasingly mobile labor force, China could retain strong comparative advantage in those manufactures for more than a decade. For example, a recent study of China's foreign trade reforms concluded that there is substantial scope for continued expansion of China's diverse manufacturing exports.[4] A large share of future exports will no doubt come from foreign-invested firms. However, it appears that many of the relatively easy one-time gains in efficiency that come from the relocation of existing export networks have already been reaped. Future diversification and upgrading of quality may be more difficult than the transfer of existing production lines. Moreover, costs are rising rapidly in Guangdong, now the manufacturing heartland of the China Circle. Congestion is increasingly driving costs upward, and wages are rising despite substantial inflows of labor. Only through a concerted and successful program of technological upgrading and removal of internal obstacles to development can a region like Guangdong stay ahead of the pressure created by past economic success.

The Attraction of the Chinese Market

The growth of the Chinese market is already playing a prominent role in shaping economic interactions both within and outside the China Circle. With China's gross domestic product poised to continue rising at about 8 percent annually for the next decade, it is inevitable that the market will be attractive to both Chinese and foreign investors. The growth of the Chinese economy is creating a new autonomous source of demand in East Asia, and this will naturally reshape development patterns. The computer industry is a good example. Whereas the China Circle is a major producer of computer products, it is not yet a major user of computer products. In 1996 the three China Circle economies accounted for 2.5 to 2.9 million shipments of personal computers, representing less than 20 percent of Asian PC demand and only one-third of Japan's 7.6 million units. Yet demand was growing more rapidly in the PRC than in any other Asian country, at 53 percent in 1996.[5] Clearly China will be substantially narrowing the gap with Japan over the next decade.

The growing market in China will shape development patterns regardless of the economic policy that China adopts in coming years. However, Chinese policy will have an important effect on the way in which that influence is exercised. Since 1992 the PRC government has been actively offering investors access to the Chinese domestic market in designated sectors, where investors are thought to bring in new technologies. This "markets for technology" policy has experienced some short-run success, as the experience of foreign investment in telecommunications equipment facilities exemplifies. But the policy carries significant dangers, because investment liberalization has outpaced trade liberalization by a large margin. In essence, foreign firms are being encouraged to invest in protected markets, and that creates hidden long-run costs. As described in chapter 3, the particular strategy of opening that began in the 1980s permitted the Chinese economy to apparently become quite "open" by many measures: exports and total trade grew rapidly as a share of GDP, and goods for export production crossed into China without tariffs and with minimal administrative barriers. This strategy allowed the export sector to race ahead of the more measured pace of liberalization of the economy as a whole. As a result, today the ordinary trading system affecting the bulk of the Chinese economy has been far less liberalized than one would expect from observing the growth of the export sector.[6] Investment inflows into the protected domestic economy therefore have several negative effects.

First, and most immediately, such investment increases allocative inefficiencies. Investment flows into protected domestic sectors without long-run competitive advantage. Some of this investment is wasted, although the effects on growth rates may not be apparent for years.[7] Second, inflows of investment without corresponding trade liberalization tend to produce upward pressure on the currency and real appreciation.

Such appreciation undermines export competitiveness and can be expected to slow export growth. Finally, following the arguments made by Michael Borrus and Dieter Ernst (chapters 5 and 7 in this volume), foreign direct investment oriented toward the protected domestic market is less likely to transfer technology as effectively as export-oriented FDI. This is ironic, because the Chinese government likes domestic market–oriented FDI precisely because it feels it can use the leverage provided by domestic market access to bargain with foreign investors for generous technology transfer. But in many cases this leverage will prove to be illusory. Foreign investors can bring showy state-of-the-art technologies to their Chinese plants, but the ultimate economic benefit for China depends on how quickly technologies are imitated and mastered by indigenous Chinese firms. Local entrepreneurs will find it harder to adapt "hothouse" technologies that are less appropriate to local economic conditions.

By contrast, a more profound opening on the part of China would create the possibility of an East Asian economic region moving toward integration, rather than simply having a successful penetration of extraregional markets. China will most likely adopt dramatic liberalization measures if terms can be agreed upon for its accession to the

World Trade Organization (WTO). That is not a sure thing, however, and the pace and nature of further liberalization of the external sector are somewhat difficult to predict. If the PRC market becomes genuinely open to trade, it will serve as the engine for a new and historically unprecedented level of Asian economic integration. Sustained Asian growth, with China as the largest source of market expansion, will ultimately convert Asia into a massive economic power that is much less dependent than now on the North American market.

Further opening of the PRC market would be highly advantageous to all the China Circle regions, including the southern provinces of the PRC. Taiwan's economy is significantly larger than that of Hong Kong, so in the long run one would expect the Taiwan economic presence on the mainland to rival or even surpass that of Hong Kong. Taiwan could conceivably emerge as a competitor to Hong Kong as an entrepôt and financial center. This emergence would be aided by Taiwan's huge investments in Southeast Asia, its big trading network, and its relatively sophisticated manufacturing base. Theoretically, there are significant complementarities between Hong Kong and Taiwan, with Hong Kong the commercial and financial giant and Taiwan the manufacturing and technology center.

Static and Dynamic Benefits of Foreign Direct Investment

Policy choices by the Beijing government about the degree of openness and the next stages of economic reform will shape the future growth trajectory of the China Circle economies. Already foreign investment is playing an extremely important role in the Chinese economies. Moreover, the potential benefits from further FDI in China are huge. Yet up till now China has reaped only a small proportion of the potential benefits from FDI. Given that the growth of output and exports has already achieved extraordinarily rapid rates, it follows that the potential for sustained growth is enormous if the obstacles to further efficiency improvement can be removed. The success of the initial stage of restructuring has simply revealed the magnitude of the potential cost advantages that can be gained by astute firms. Whole new areas of potentially profitable activity have been disclosed by the success of the first stage of integration. Even with the limitations of the current environment, a greatly expanded market is emerging in the

China Circle for intermediate and investment goods, some of which will displace current PRC imports. Many foreign-invested enterprises are already seeking sources of intermediate goods to replace the current imported supplies. Some will come from existing PRC firms, and some will come from additional foreign investment, which is expanding in, for example, production of rubber and plastic raw materials.[8] At present, linkages to upstream producers are being created by additional foreign investment; if further reforms succeed, domestic firms will also play a large role.

The benefits from FDI can be divided into static and dynamic benefits. Static benefits accrue primarily from the relocation of production chains to take advantage of lower costs, especially lower labor costs. Not surprisingly, foreign investors in China have been able to reap the lion's share of the static benefits. Hong Kong and Taiwanese businesses possessed the crucial strategic intelligence about world markets and marketing, packaging and sales, and production and transport coordination. They also had valuable reputational assets: importers know them to be quick, reliable, high-quality producers. Moreover, they usually remained in full control of the total chain of transactions. Indeed, the opening of the mainland allowed Hong Kong and Taiwanese investors to squeeze additional value out of assets that might otherwise have become obsolete: Investors from Taiwan and Hong Kong often contributed used, partially or wholly depreciated, machinery to joint ventures on the mainland. Such machinery would have had little value in Taiwan or Hong Kong, where increasing labor costs and tighter pollution standards were making them obsolete.[9] Conversely, the PRC was in competition with other low-wage locales. Competition for investment and the initial distribution of assets inevitably implied that most of the static gains would accrue to the non-PRC partners.

In the long run the dynamic benefits to investment are more important. Dynamic benefits accrue primarily through the learning process: they accumulate when individuals imitate and learn to begin new activities and to make existing activities better. Whatever the static benefits of FDI to Hong Kong and Taiwan, it was the exploitation of the dynamic benefits that was truly impressive. Everywhere one looks in the economic history of Taiwan and Hong Kong, one finds a process of rapid imitation, of ceaseless movement into new activities, and of uninterrupted acquiring of new capabilities. Borrus, Chin Chung, and Ernst have all documented different aspects of this process in the electronics industry. Clearly success in this area was related not only to good human resources (relatively high education levels in particular) but also to a fluid, competitive environment in which the returns to inventive activity were high and the barriers to entry low. Moreover, Ernst argues that capabilities are transferred more rapidly when transnational networks are "open," that is, when local affiliates are given decision-making authority, when locals participate in management, when development and research activities take place locally, and when indigenous companies are encouraged to supply inputs and enter into subcontracting relations. Both Ernst and Borrus note that the openness of networks is strongly conditioned by firm strategy. Export-oriented firms in competitive fields have a much stronger motivation to build local supply capacity than do investors targeting protected local markets. Therefore, there is substantial reason to believe that the past experience of Taiwan and Hong Kong could be replicated on the mainland, and open firm networks could build strong domestic capabilities.

The result so far seems fairly disappointing, however. During the mid-1980s the semireformed Chinese economy was unable to provide a reliable and timely supply of inputs, so the Chinese government encouraged a policy of "both ends outside," permitting foreign investors to source raw materials and components outside China, without import duties, as long as final products were also exported. As a result, supply links to the rest of the PRC economy were minimized, and opportunities for spillover learning limited. Local producers do not have much knowledge of final markets and are not really in a position to gain that knowledge, since the final stages of packaging, shipping, and marketing are usually performed off-site, in the home region of the investing firm. Thus the spillovers of knowledge into related activities are below potential, and China remains concentrated in low-wage, low-skill activity.

Other rigidities in the PRC economic system inhibit the capturing of dynamic benefits. Factor markets in the PRC continue to be highly distorted despite fifteen years of economic reform. The financial system essentially rations credit to designated users. Land mortgages are not available. Foreign investors are expected to bring their own funds and thus do not have access to a well-functioning banking system. This affects ordinary trade financing as well as the nurturing of subcontractors. Furthermore, the mainland's macroeconomic policy seems to be in perpetual disequilibrium, with the credit supply either expanding rapidly or going through a periodic credit crunch. These conditions slow the growth of links between foreign-invested firms and the domestic economy and perpetuate the existing system.

Most important, though, there are still significant limitations to the ability of entrepreneurs in China to personally capture the benefits of economic spillovers. There are no longer strong obstacles to creating a small private firm in China, but there are formidable obstacles to the entrepreneur who wishes to build a large firm, particularly one that crosses geographic boundaries. Property rights are poorly protected, especially when interregional transactions are involved. Private firms find it nearly impossible to get direct import and export rights. State-owned firms often have powerful patrons that restrict the competitive opportunities of private firms that seek to grow. For all these reasons, both Chung and Jean François Huchet (chapter 8) are skeptical of the Chinese ability to respond fully to the opportunities presented by foreign investment inflows. The creative substratum of entrepreneurial small firms is simply not present. Time may be a factor: there has not been enough time to build up a network of small firms with substantial technical and managerial capabilities. The rigidities that prevent China from realizing the potential dynamic gains from investment come from the Chinese economic system itself, not from the nature of the foreign investment.

Potential Political Conflict

In purely economic terms the return of Hong Kong to the PRC on July 1, 1997, is of relatively little importance. As has been often noted, Hong Kong has already become closely integrated with the Guangdong economy, and none of the political changes in themselves will have a significant impact on the economic rules of the game. However, Hong Kong's retrocession creates new kinds of political risk, particularly when combined with the uncertainty caused by the transition in Beijing to the post–Deng Xiaoping leadership. During 1995 and 1996 the Beijing government took an extremely tough line toward the Hong Kong government and the Hong Kong democratic opposition, so that the Chinese government is now positioned to adopt a more relaxed attitude in the months after July 1, 1997, if it so chooses. One can therefore expect the actual turnover to be relatively smooth. But in subsequent months and years the Beijing government will face repeated policy choices, and occasional challenges to its authority, in Hong Kong. It is impossible to say how those choices will be handled in all cases; there is thus an irreducible addition to overall political risk associated with the handover, no matter how sanguine one is about the intentions of the Beijing government. Such uncertainty will inhibit some investment in the China Circle, particularly when combined with the difficulty in discerning the PRC government's attitude toward relations with Taiwan.

After a period of saber rattling in early 1996, relations between Taiwan and the PRC had settled into a more "normal" pattern by the end of 1996. It is striking that the overall bargaining relationship between Taiwan and the PRC was not fundamentally altered by the events of 1995–96. Taiwan failed to qualitatively change its peculiar status as a state almost entirely without diplomatic recognition, while the PRC completely failed to achieve any of its short-term objectives through military intimidation.[10] Instead, suspicion and distrust between the two sides intensified, and the relationship was maintained within the same basic framework as before. Since the period of crisis some signs of progress have been evident, including the initiation of direct shipping between the PRC and Taiwan, albeit still under restrictive conditions.

In fact, there are several things that might improve the bargaining position of Taiwan over the next few years. As the nature of investment in the PRC changes, the country is shifting from small, dispersed labor-intensive manufacturing projects that are hard to monitor toward larger raw material and infrastructure projects that are much more visible to the Taiwanese government. Thus the government may have a greater ability to restrict Taiwanese investment, thereby gaining leverage over the mainland. Taiwan's position has already been improved and may be further improved by its successful transition to a democratic government. Moreover, continued change in official attitudes toward the PRC may help Taiwan in the long run. The most obvious recent manifestation of that change was the issuing by the Taiwanese government of a document entitled "Seeing through the One China Policy" (of the PRC government). This document laid out the Taiwan position in clear and nonideological language. Finally, as Taiwan through the end of the 1990s takes delivery of military hardware previously purchased from the United States and France, the military threat from the PRC will recede to the smallest level in recent and foreseeable times. In fact, Taiwan will enjoy a modest military superiority over the PRC for at least a few years.

But weighed against those factors is the growing economic weight of the PRC and the increasing involvement of Taiwan in that economy. Already there are questions whether the Taiwan government is really in a position to prevent large Taiwanese corporations from investing in large mainland projects, even when they are highly visible. As of early 1997 the giant Formosa Plastics Company was declaring its intention to proceed with a large power plant in Fujian despite the vehement opposition of Taiwan's president, Lee Tenghui.

More broadly, although ample tension and grounds for conflict exist, it is striking how little margin of error the respective governments really have. For Beijing, policies that seriously undermine the prosperity of Hong Kong will almost certainly create major problems for whichever leader is seen to be responsible. Such a leader would be vulnerable to charges that he had bungled the historic opportunity to reunite China on peaceful terms and had ruined the opportunity to attract Taiwan back to the mainland. It is unlikely that a leader could survive in power under those conditions. Conversely, in Taiwan, a democratically elected leader could not really expect to retain power if he mangled the prosperity of Taiwan, which is increasingly dependent on at least a modus vivendi with the mainland. Thus while it is impossible to be sanguine that policymaking in Beijing will be wise, it is at least reassuring to expect that it will be fairly accountable. Leaders associated with good policies will have an advantage in the leadership competition, and leaders associated with foolish policies are likely to be discarded quickly. As a result, there are grounds for some optimism. The most likely outcome may be a continuation of tension and a complex, constantly evolving negotiation process, but one that, with reasonable luck, can be kept from exploding into violence and conflict and that can contribute to a more prosperous future.

The Future of the Chinese Economies

The East Asian economies are all engaged in an extraordinarily dynamic process of economic growth. The metaphor of flying geese has been popularly used to describe a process in which all the East Asian economies develop rapidly but in which the relative positions of each do not change. In this metaphor, Japan remains the head goose, followed by the four newly industrializing countries—Korea, Taiwan, Singapore, and Hong Kong—with China and the ASEAN countries next in line. The experience of the China Circle gives some indication that these relative positions do have the potential to change. Within the dynamic East Asian region as a whole, the complementary capabilities of different parts of the China Circle might enable an even more rapid flight for some China Circle members than for other economies in East Asia. The capabilities of the Taiwan computer industry may propel it to a position near the frontier of world computer manufacturing (see chapter 6). The capabilites of China manufacturing combined with the potential attraction of the China market may push China ahead of its competitors in ASEAN. Yet it is also clear that for these changes to take place, further policy reforms are necessary that go beyond what is currently emanating from Beijing. But if past experience is any guide, policymaking will eventually adapt to the opportunities presented to it and propel the China Circle into a new era of development.

The future of the China Circle will be determined by the evolution of policy in Beijing. In an optimistic scenario, China would move decisively toward a more open economy, with the creation of strong market-based institutions and property rights. In that case, the past growth in the China Circle would feed into a broad-based process of development, bringing China into the mainstream of East Asian growth. In a pessimistic scenario, China would linger in a kind of transitional halfway house, maintaining an economic system that basically reflects a market economy, but one ridden with market distortions, with significant protection for the domestic market, and with incentives undermined by insecure property rights and the threat of arbitrary action by government officials.

In the short run there is likely to be some hesitation and wasted time: growth slowed in the China Circle in 1996, and slower growth will probably prevail through 1997 and perhaps 1998. But in the longer run Chinese policymakers will be forced to come to grips with the fundamental policy issues and grapple with the need to integrate China into the world economy. Chinese policymaking has often been inconsistent over the past twenty years, but Chinese policymakers have generally been able to respond effectively when faced with clear and unambiguous problems or challenges. In the case of trade and investment policies related to the China Circle, the shortcomings of continued policies of domestic protection, a dualistic trade regime, and an excessive need for political and economic control are likely to become increasingly evident. Under such circumstances, it is reasonable to expect that China's leaders will move, within a few years, to enact a further degree of openness and reform in the Chinese economy. That in turn would set the stage for a new phase of growth in the China Circle, but a growth that would see the China Circle lose its special identity and emerge as only one—but one of the most dynamic—of the economic centers along the East Asian edge between Tokyo and Singapore.

Notes

1. Barry Naughton, "China's Emergence and Prospects as a Trading Nation," *Brookings Papers on Economic Activity, 2 :1996,* pp. 316–20.

2. Economist Intelligence Unit, "Sayonara Dalian: Japan's Waning Affair with the North-East's Most Prosperous City," *Business China,* March 17, 1997, pp. 3–4.

3. Computer-related products refers to standard international trade classification category 75, office machines and automatic data processing machines; garments to SITC category 84; and miscellaneous manufactures to SITC category 89. *China Customs Statistics,* 1989:4, p. 15; December 1995, p. 8; December 1996, pp. 8–9.

4. World Bank, *China: Foreign Trade Reform* (Washington, 1994), pp. 158–66.

5. Jon Skillings, "In Asia, Japan Buys More PC's, but China Buys More Quickly," IDG *China Market News,* February 18, 1997 [via World Wide Web, at http://www.IDGChina.com/idgnews.htm].

6. Barry Naughton, "China's Emergence and Prospects."

7. Those sectors enjoy higher prices owing to tariff barriers. As a result, their output has a large weight in total GDP, and their growth may boost short-run GDP growth rates. However, growth will suffer in the long run if it does not prove possible to boost productivity in protected sectors toward internationally competitive levels.

8. "Synthetic Rubber Factory Planner for Mainland," *Free China Journal,* July 29, 1994, p. 3.

9. Sumner La Croix and Yibo Xu, "Political Uncertainty and Taiwan's Investment in Xiamen's Special Economic Zone," in Sumner La Croix, Michael Plummer, and Keun Lee, eds., *Emerging Patterns of East Asian Investment in China: From Korea, Taiwan and Hong Kong* (M. E. Sharpe, 1995), pp. 123–41; and Richard Pomfret, "Taiwan's Involvement in Jiangsu Province: Some Evidence from Joint-Venture Case Studies," in ibid., pp. 167–78. Similarly, the intangible assets Taiwan and Hong Kong businesses held were in danger of becoming obsolete. Changing cost structures would inevitably have pushed Taiwan and Hong Kong out of the mostly unskilled labor-intensive sectors, where they had substantial reputational advantages. However, the opportunity to transfer production to the mainland allowed businessmen to continue to earn value out of those reputational advantages and real skills, which might otherwise have become irrelevant as production moved to more distant low-wage locales.

10. The one possible exception is that the PRC may have achieved some bargaining advantage from demonstrating that its position on Taiwan was so strongly and irrevocably held that it could be expected to engage in behavior that was destructive both to itself and to others if its core interests were threatened.

Abbreviations

APEC	Asia-Pacific Economic Cooperation
ASEAN	Association of Southeast Asian Nations
ASICs	application-specific integrated circuits
CEMI	Chinese-European Management Institute
CITIC	State Council's international investment arm
CTT	Posts and Telephone Administration (Macau)
DGT	Directorate-General of Telecommunications
EC	European Community
EU	European Union
FDI	foreign direct investment
FIEs	foreign invested enterprises
FTC	foreign trade corporations
GSC	Greater South China
GSP	generalized system of preferences
HKTC	Hong Kong Telephone Company or
IC	integrated circuit
IPLCs	international private leased circuits
IPOs	International Procurement Offices
ITRI	Industrial Technology Research Institute
JACTIM	Japanese Chamber of Trade and Industry in Malaysia
KLSE	Kuala Lumpur stock exchange
KMT	Kuomintang
MPT	Ministry of Post and Telecommunication
NIEs	newly industrializing economies

OC	overseas China
ODM	original design manufacturing
OEM	original equipment manufacturing
OFTA	Office of the Telecommunications Authority
PC	personal computer
POTS	plain old telephone services
PRC	People's Republic of China
PSTN	public switched telephone network
PTAs	Posts and Telecommunications Administrations
SEZ	special economic zones
VLSI	very large-scale integration circuits

The Contributors

Michael Borrus
Berkeley Roundtable on the International Economy
University of California, Berkeley

Chin Chung
Associate Research Fellow
Chung-hua Institution for Economic Research, Taipei

Dieter Ernst
Professor, Department of Development and Planning and Department of Business Studies
Aalborg University, Denmark
Senior Research Fellow
Berkeley Roundtable on the International Economy
University of California, Berkeley

Jean-François Huchet
Research Fellow
Maison Franco-Japonaise, Tokyo

Barry Naughton
Graduate School of International Relations and Pacific Studies
University of California, San Diego

Yun-wing Sung
Department of Economics
Chinese University of Hong Kong

Tan Kong Yam
Head, Department of Business Policy
Faculty of Business Administration
National University of Singapore

The University of California Institute on Global Conflict and Cooperation

The University of California Institute on Global Conflict and Cooperation (IGCC) was founded in 1983 as a multicampus research unit serving the entire University of California system. The institute's purpose is to study the causes of international conflict and the opportunities for resolving it through international cooperation. During IGCC's first five years research focused largely on the issue of averting nuclear war through arms control and confidence-building measures between the superpowers. Since then the research program has diversified to encompass several broad areas of inquiry: regional relations, international environmental policy, international relations theory, and the domestic sources of foreign policy.

IGCC serves as a liaison between the academic and policy communities, injecting fresh ideas into the policy process, establishing the intellectual foundations for effective policymaking in the postwar environment and providing opportunities and incentives for UC faculty and students to become involved in international policy debates. Scholars, researchers, government officials, and journalists from the United States and abroad participate in all IGCC projects, and IGCC's publications—books, policy papers, and a semiannual newsletter—are widely distributed to individuals and institutions around the world.

In addition to projects undertaken by the central office at UC San Diego, IGCC supports research, instructional programs, and public education throughout the UC system. The institute receives financial support from the Regents of the University of California and the state of California and has been awarded grants by such foundations as Ford, William and Flora Hewlett, John D. and Catherine T. MacArthur, Rockefeller, Sloan, W. Alton Jones, Ploughshares, the Carnegie Corporation, the Rockefeller Brothers Fund, the United States Institute of Peace, and The Pew Charitable Trusts.

Susan L. Shirk, appointed in July 1997 as the deputy assistant secretary, Bureau of East Asian and Pacific Affairs, U.S. Department of State, is a professor in UC San Diego's Graduate School of International Relations and Pacific Studies and in the Department of Political Science. She was IGCC's director from 1991 to June 1997. Former directors of the institute include John Gerard Ruggie (1989–91) and founder Herbert F. York (1983–89).

Index